Peacetime Uses of Outer Space

PEACETIME USES OF OUTER SPACE

LLOYD V. BERKNER

JAMES H. DOOLITTLE

FREDERICK R. KAPPEL

VICE ADM. JOHN T. HAYWARD

LESTON FANEUF

LEO GOLDBERG

JOSEPH KAPLAN

MORRIS NEIBURGER

WILLARD F. LIBBY

OVERTON BROOKS

RALPH J. CORDINER

BRIG. GEN. DON FLICKINGER

EDWARD TELLER

Edited by
SIMON RAMO
Executive Vice President
Thompson Ramo Wooldridge Inc.

GREENWOOD PRESS, PUBLISHERS
WESTPORT, CONNECTICUT

Library of Congress Cataloging in Publication Data

Ramo, Simon, ed.
    Peacetime uses of outer space.

    Reprint of the 1961 ed. published by McGraw-Hill,
New York.
    Includes index.
    1. Astronautics. 2. Astronautics and civilization.
3. Outer space--Exploration. I. Berkner, Lloyd Viel,
1905-   II. Title.
[TL790.R27 1977]   333.9'4   76-52430
ISBN 0-8371-9368-0

PEACETIME USES OF OUTER SPACE

Originally published in 1961 by McGraw-Hill Book Company, Inc.,
New York

Reprinted with the permission of McGraw-Hill Book Company

Reprinted in 1977 by Greenwood Press, Inc.

Library of Congress Catalog Card Number   76-52430

ISBN 0-8371-9368-0

Printed in the United States of America

Preface

THE INCREASING importance of science and technology in shaping our personal lives requires a deeper appreciation by all citizens of the potential effects upon our civilization of the continuing advances in these fields. And of all the new advances, none is evolving more rapidly nor is more fraught with possibilities for radically changing the future than our ventures into outer space, which have barely begun.

It was for these reasons that the University of California made the decision in mid-1959 to organize and sponsor a series of lectures intended for the general public to be given by prominent authorities on various aspects of "The Peacetime Uses of Outer Space." Through the able efforts of L. M. K. Boelter, Chairman of the Department and Dean, College of Engineering, Los Angeles; Clifford Bell, Head, Physical Sciences Extension, Southern Area; and John C. Dillon, Head, Engineering Extension, Southern Area, arrangements were made for a series of twelve lectures. The speakers chosen were not only recognized experts in their particular subject matter but were distinguished for their skill in presenting information to a large, nonspecialist audience. The lectures were given during the period March 23 to June 29, 1960. Each lecture constitutes a chapter of this book, although their sequence has been rearranged.

The book, as was the lecture series, is addressed to the intelligent, interested layman. Its purpose is not only to provide information about the coming space age, but also, hopefully, to convey a broad feeling of how developments in science and technology are having an increasingly greater impact upon our everyday life. The transition we must make to the complex, fast-paced, highly technological civilization toward which the world is headed can be orderly or chaotic, largely as we succeed or fail in understanding the nature of such underlying forces. Outer space has already become both a symbolic and a substantive example of scientific and technological change.

The rapidity of the advances in our space programs has also raised a problem in the editing of this book. In a number of instances, new and significant developments occurred within a few weeks after a particular lecture was given. It has therefore seemed desirable in some instances to bring the subject matter up to date within the limits imposed by the publishing schedule. Accordingly, some of the chapters have been revised to take into account certain important developments through October, 1960.

The careful reader will notice occasional differences between the information and opinions given by one author as compared with another. This, of course, is a normal state of affairs when one is dealing with the frontiers of science and technology; such differences cannot usually be resolved until further advances have taken place. The reader may also find some degree of redundancy between certain chapters. This, again, has deliberately been left unchanged. The stature of the authors and the breadth of the topics being covered make it worthwhile to consider each author's views separately.

I should like to thank M. F. Thorne of the Thompson Ramo Wooldridge Corporate Staff without whose help I could not have discharged my duties in a timely fashion and who assisted me in every aspect from arrangement-making and communication to editing.

Finally, I want to join the University of California in expressing appreciation to each lecturer-author, who in every case had to find time in an exceedingly busy schedule in order to participate in the series.

SIMON RAMO

Contents

1

Space Research—A Permanent
Peacetime Activity

LLOYD V. BERKNER
PRESIDENT
GRADUATE RESEARCH CENTER OF THE SOUTHWEST

Blackstone Studios

L. V. Berkner is President of the Graduate Research Center of the Southwest, Dallas, Texas, an institution to advance graduate education in the Southwestern United States. He is active in organizations of international scientific activities as Retiring President of the International Council of Scientific Unions and Retiring Vice President of the Special Committee for the International Geophysical Year; Past President of the International Scientific Radio Union. He is Treasurer of the National Academy of Sciences, Chairman of the Space Science Board of the National Academy of Sciences, President of the American Geophysical Union, and President of the Institute of Radio Engineers. He is also a Consultant, President's Science Advisory Committee. B.S. (Electrical Engineering), University of Minnesota, 1927 (Distinguished Alumni Award 1952). Honors: D.Sc., Brooklyn Polytechnic Institute; Ph.D., Uppsala University, Sweden, 1956; D.Sc., University of Calcutta, India, 1957; D.Sc., Dartmouth College, 1958; D.Sc., University of Notre Dame, 1958; D.Sc., Columbia University, 1959; LL.D., University of Edinburgh, Scotland, 1959; D.Sc., University of Rochester, 1960. Holds U.S. Special Congressional Gold Medal and numerous other awards and commendations.

TO UNDERSTAND the place of research in generating an intelligent space program, I will ask you to imagine with me our space situation in the decade of 1970 to 1980. What vehicles will we have; what kinds of exploration will conceivably be within our reach?

By 1965, the Nova engine will have reached the advanced test stage, providing us with a basic thrust of perhaps 1 million pounds. Further development to an advanced configuration will lead to increased thrust of this or a similar chemical engine—perhaps 2 million pounds by 1970— and repeated tests will have made it reliable. At the same time, we will have had several flights with the Saturn vehicle in the 1965 to 1970 era, using several clustered rockets of the type already used for space flights. So the use of clustered boosters for the first stage of launching our space vehicles will be familiar, and the reliability of this type of vehicle will have been established.

Then, by 1970 we will have started to think of clustering as many as seven Nova engines to give us perhaps 15 million pounds of initial thrust. From this we can estimate an ultimate maximum space payload between 1970 and 1975 of 250,000 to 400,000 pounds—let us say 200 tons maximum. Perhaps improvements in fuel will raise this to 250 tons, but this will be the approximate limit.

To put this in terms we can visualize, let us think of the Boeing 707 as the standard unit of weight and size; this would provide for about two units of our 707 as the space payload.

Now, if we go to a planet and want to come back, at least one-quarter of this load must go for fuel to permit landing, and one-half of it must go for the return rocket and its re-entry provisions upon its return. Then we need a large electric power supply and radio transmitters and computers to reduce data. To this we must add the necessary food, water, and oxygen to sustain two or three or four men for perhaps a couple of years, as well as the weight of shielding against radiation and other ecological aids. Finally, we will have scientific gear for very special problems of exploration of the planet.

2

When you start to add up the weights, you use up your two 707's in a hurry and conclude that really significant manned space exploration to a planet will perhaps just become possible in the 1970 to 1980 era. Whether it will be possible depends upon two factors:

1. The development and test of reliable rocket and vehicle hardware

2. The origination of scientific apparatus of a suitable kind to do significant scientific exploration

When you add man, then the whole problem of man's ecology, his relationship to the environment of the vehicle and the planet, must be worked out.

The objective of such an operation would probably be the first explorations of the surface of Mars. Why not Venus, you ask? Well, recent measurements show Venus to be very hot—400 to 500° F. Certainly, we will improve these measurements and perhaps even revise our estimates radically, but at the moment, Mars seems to be the least formidable of the planets for a manned landing. Moreover, we know that Mars has some atmosphere, probably of mostly inert gases, and it does have some water vapor. There are winds, and a form of meteorology on Mars, though there is no evidence of any free oxygen. The temperature of the planet is not far below that of the earth. So the atmosphere of Mars and its greater distance from the sun would provide some protection from the lethal radiations of ultraviolet and X rays and the very lethal clouds of particles emanating from solar eruptions. Moreover, there is strong evidence of some form of life on Mars, since the spectra of carbohydrate molecules are found in its reflected light. Similar spectra are characteristic of life on the earth.

But whether the objective is Mars or Venus, the difficulties and problems are of comparable magnitude.

The cost of such an expedition is large—one can estimate it at perhaps 250 million to 1 billion dollars per flight. The supporting activities will be of equivalent cost. So we cannot expect a flight of this kind more than once each two or three or four years, and even then our total space bill with auxiliary activities will amount to 2 or 3 billion dollars per year. This will include perhaps 1 billion on operations, 1 billion on engineering and hardware, ½ billion on development, and ½ billion on research leading to, and effectuating, sound scientific programs.

This is a large amount of money, and you may well ask whether space is a permanent hole down which the taxpayers' money will continuously flow. I think not. Our research and operations already show the way toward large financial returns. Later in this book you will read more about these returns, but let me mention a few:

1. Military
2. Commercial
3. Scientific

The use of satellites holds the promise of advanced forms of military reconnaissance and communications intelligence. Early warning of missile attack may become feasible. The reconnaissance satellite may be the means of removing doubts and calming tensions that might otherwise lead to war. Such satellites will certainly provide an important adjunct to inspection that may permit a safe agreement on arms limitation. So the military applications of an advanced space program may be worth billions.

Likewise, the provision of new forms of satellite relay communications is destined to multiply our available long-distance communications by a factor of perhaps 10 thousand. Our present ionosphere supports a communication band of about twenty megacycles. This band limits the total amount of long-distance communications. A single satellite may carry a relay band one hundred times as broad. In 1960 we saw the first experiments of bouncing radio waves from a large metallic sphere over long distances. Soon this will be followed by the active satellite that receives, amplifies, and relays the communication. In a few years such satellites will be in hovering orbits, for a satellite launched to 22,400 miles above the surface, and eastward along the equator, goes round the earth at the same rate the earth rotates. So it hovers over a single meridian. Such a satellite, with very little energy to stabilize it, provides a permanent relay station overhead. Beyond the hovering relay satellite, we can foresee even more sophisticated means of space radio relay.

So space will multiply our communications industry by many times. It offers the probability of world-wide telephone dialing; of creating new industry made possible by cheap, fast, and unlimited communication. Again, the value will be measured in many billions of dollars annually.

Similarly, we can expect that the application of space techniques to earth travel at hypersonic speeds will have commercial uses also valued in the billions of dollars.

Turning to the rewards from scientific progress, sparked by space, the value to man is more difficult to evaluate. If we knew ahead of time what an experiment would yield, we would not have to do it. But we do know that our advance in knowledge always yields useful results—indeed, almost all the new industry of the present century has grown from recent scientific progress. So we are certain that space research will provide valuable returns in industry and employment and human welfare, even though at the moment we cannot pinpoint the exact gains. Already, the beginnings of space exploration have given us the new opportunities in military power and commercial communication that I have just cited.

Certainly advances in our knowledge of meteorology will come in the near future. Satellites are already measuring the input solar energy and the outward radiation of heat from the earth, so we are beginning to map the heat distribution that motivates the winds and the storms. In 1960 we took the first step in televising the cloud cover over the earth to map the ever-changing weather system from the signatures of the cloud tops. Pictures of the weather over the whole Atlantic taken from rockets 700 miles overhead also have given new clues to storm intensification or dissipation. Soon we will know how many storms exist over the earth at one time, how they are growing, moving, and disappearing. Who can doubt that our knowledge of weather will change mightily as the space program progresses.

The new astronomy from space allows us to view the sun and the universe in the full range of chromaticity, from the longest to the shortest wavelengths. Already photos of the sun have been taken from rockets in ultraviolet light that give new clues to its physical structure and dynamics. Our knowledge of the nature of matter, its origin and its destruction, will certainly be advanced by space astronomy.

In space itself, our science has already found new surprises, and many others are doubtless in store. For 100 years, man had puzzled about the connection between storms on the sun, with their violent solar chromospheric eruptions, and the auroras and the storms in the earth's magnetism that followed them. Then, in one great discovery by Prof. James Van Allen of the radiation belts encompassing the earth, the facts of the mystery began to fall in place. Beginning with Explorer IV and Vanguard II, and going on to Explorer VII and Pioneer V, satellites have made it possible to map these great belts in their configuration around the earth. As the sun erupts, dense clouds of very energetic protons, the elementary nucleii of hydrogen, course toward earth. Their density is so great as to be lethal to man in interplanetary space unless he is adequately shielded from radiation levels that sometimes surpass 10 rem per hour even behind modest shielding. (A level of 400 rem is fatal.) Some of these particles are captured in the Van Allen belts and steered toward the poles where they plunge into the high atmosphere to produce the auroras. The disintegration of atoms of the atmosphere upon collision with cosmic rays provides particles that enhance the electrification of parts of these radiation belts.

So, we are on the threshold of understanding phenomena of space by acquiring new knowledge that we did not even suspect before. Clearly, these intense radiations must be mapped and understood in great detail before we can send man into interplanetary space with adequate protection. Until this is accomplished, our manned space flights must be confined to satellites orbiting within a couple of hundred miles of the earth, preferably near the equator.

In ranging through space we can study weak fields of gravitation and magnetism, and perhaps learn to understand the character and significance of the magnetohydrodynamic waves that have been postulated by theoreticians. Who knows what their significance might be to our earthly environment? But certainly that knowledge will greatly advance our understanding of the physical basis and origin of our universe and the planetary system that is our home.

Returning to the earth itself, geodetic satellites with powerful flashing lights and special radios will provide mapping of the earth that is accurate to perhaps 50 feet. We can then test such ideas as continental drift without waiting a millenium. We can operate satellites within the ionosphere itself to test the high atmosphere in the very regions where the radio waves are now reflected for long-distance transmissions. We can go out into the very fringes of the atmosphere to examine those strange phenomena on very low radio frequencies known as whistlers to learn in detail of their origin and means of propagation.

We could go on and on describing the immense range of scientific experiments that space exploration has now enabled us to contemplate.

New clues to nature's processes, formerly beyond our reach, can now be followed to expand the knowledge of our environment. The new scientific progress made possible by space exploration is certain to pay off in many important ways.

In speaking of the rewards of space activity, we cannot ignore two others; these are the political and the spiritual. The United States with her leadership in advancing man's welfare—a leadership triggered by a superb science and technology—has become the standard that must be surpassed by any other nation if that nation is also to claim the distinction of leadership. Therefore, such emergent societies as Russia and China aspire vigorously to surpass the United States to demonstrate the superiority of their ideology and social system. They have set this as their objective unambiguously.

Escape into space—the exploration of the heavenly bodies around us—is a deep-seated aspiration of all mankind. That escape was reserved to man's early gods, and it is closely identified with civilization's early origins in mysticism, folklore, and religion. At every intellectual level man longs to know the nature of other bodies around him, so astronomy was among his earliest sciences. Man prizes this idea of escape from the earth to the universe as the highest symbol of progress. Therefore, the nation that can capture and hold that symbol will carry the banner of world leadership. Consequently, leadership in space exploration has a real political meaning. Failure in that leadership means inevitably falling into the status of a second-class nation with the heavy costs to our way of free enterprise which subjugation to others would involve.

George V. Allen, Director of the United States Information Agency, in January, 1960, said before the Congress:

> . . . regardless of how Americans may feel about it, this country is in a space race as far as world opinion is concerned. It is a race we can't avoid and that we might as well accept. . . . As a result of Soviet space successes of the past two years, there has already been a significant change in world opinion about Soviet science, technology, military power and general standing. . . . The Soviet Union is able to challenge America in all those fields, including even production.

His words were echoed by Under Secretary of State Livingston T. Merchant:

> By being first to achieve success in space flight, the Soviet Union has reaped great prestige. Continuing achievements have made this gain an enduring one. It has become apparent to all that the Soviet Union is capable, when it chooses to concentrate its efforts, of pioneering work in advanced and difficult fields of science and technology . . . [it] is not limited to following and imitating the achievements of Western science and technology. . . .
>
> As one would expect, Soviet propaganda has with some success capitalized on the technological achievements of the Soviet Union by attempting to present an image of preeminent achievement, not merely in science and technology, but across the board, including military power.
>
> It would be wrong and dangerous to discount either the achievement or the impact of that achievement in the minds of peoples all over the world.

This is the sober assessment, 2½ years after Sputnik, by officials who are in daily contact with international affairs. When acquired, prestige can only be lost, and as a leader of the free world we cannot afford to desert our responsibilities.

So we must not only balance the costs of our space program against the expected dollar returns in the military, commercial, and scientific sense, but we must also calculate the immense costs if we fail to undertake a significant program to retain the symbol of leadership. These costs may range from simple loss of commerce to the whole subjugation of our free system and the encroachment of an arrogant and powerful system of dictatorship that has gathered world power under its symbol of leadership.

However, the spiritual values of a sound space program go far beyond this. The objectives of the East and the West are the same in satisfying man's aspirations to explore the universe. This common purpose and objective of all peoples may become a binding influence—an overweening area of common cultural understanding—which can bring them together in agreements on an ever-widening base. As Reinhold Niebuhr has remarked, law among nations is ineffective unless there are threads of common understanding and culture that bind nations together. Space science and exploration can become such a thread.

Above all, space exploration can satisfy man's curiosity about the universe in which he lives—a curiosity whose pursuit has given him his civilization. I am reminded of the words of the famous explorer, Fridtjof Nansen, chiseled in the oak above the entrance to the great library at Carleton College:

> The history of the human race is a continuous struggle from darkness toward light. It is therefore of no purpose to discuss the use of knowledge—man wants to know and when he ceases to do so he is no longer man.

As you can see, the rewards of space exploration are great—certainly in dollars—but even more in terms of our political and spiritual values.

How do we acquire these rewards? At the time this is being written, "space" is hardly 2½ years old, dating from the launching of the first satellite on October 4, 1957. In the intervening time we have hurriedly assembled the National Aeronautics and Space Administration and initiated a substantial space program. Our appropriations this year will approximate 1.1 billion dollars.

The surprising thing is that the program has been as effective as it has, starting from scratch when we did. It is certainly true that the Soviet Union surpassed us at the start, and will continue to have larger vehicles for some years to come. But in spite of this advantage, our science has been absolutely first-rate. Because of the scientific planning and the development of advanced scientific instruments for the Vanguard satellites in the IGY, United States scientists were ready for a variety of well-prepared experiments. As fast as the vehicles have become available, these experiments have been done. Consequently, in spite of the Russian advantage in larger vehicles, the United States needs make no apology for the quality of its scientific work—indeed, in this department we now have the leadership in most respects. The problem is to keep it that way and to improve the situation. This means that our scientific preparation for space must always be far ahead of the availability of vehicles if each space vehicle is to achieve its maximum potential.

Here lies the greatest challenge of our space program. Underlying the planning for each flight must be a strong, continuing, and very basic scientific program that advances the preparatory science to the point where each vehicle can carry on an optimum program of scientific exploration.

Without doubt, each payload into space must be carefully planned as a unit for a long time in advance. That means early perfection of scientific apparatus and careful engineering of each part into the space vehicle with all its auxiliary needs for power, temperature control, stability, orbit control, and the like. To accomplish this needed engineering design and development, we have two excellent laboratories in the government's National Aeronautics and Space Administration, the George C. Marshall Space Flight Center and the Goddard Space Flight Center, which are well organized to handle these engineering matters. The Jet Propulsion Laboratory at the California Institute of Technology, also NASA supported, together with such industrial laboratories as General Electric Space Center and Space Technology Laboratories, Inc., are well organized and equipped to carry out space engineering in the area of private organizations. The activities of these laboratories have a major part in the space program, and they are making and will make in the future great and essential contributions.

But we have to be careful that emphasis on the engineering of the payloads and on the practical problems associated with each launching does not encourage neglect of the basic scientific program that is necessary and essential to fulfill the real purpose of a sound space program. This fundamental objective must be a basic, broad, and unfortunately expensive program of research that is directed toward maximizing space objectives but is never tied to a specific launching. Only such a program will provide the advanced and significant ideas and conceptual devices from which subsequent launchings can be organized.

This absence of a foundation program of research is not dangerous for the time being when science is ahead of the vehicle availability. But this advantage will soon be lost as vehicles become more reliable and numerous in the next year or two. Therefore, the danger of neglect of basic science is very real when we think in terms of the 1970 to 1980 period, and the scientific foundation that will be needed for that era. Unless we very quickly undertake an immense program of very basic science and technology oriented toward space exploration, our present advantage of scientific leadership in space science will become sterile and be lost. Then we will be behind in *both* vehicles and science.

Well, you say, what is the matter with pressing our vehicle program so that we can do a few stunts like hitting the moon—and include whatever scientific gear happens to be lying around. Wouldn't that be enough? The answer is a clear "no." Men everywhere are quick to detect a phony. Already the U.S. has taken the edge off the Russian precedence in vehicles by the excellence of U.S. science. The real objective in space is to find out about it, and the nation that does will ultimately steal the show.

Each of our vehicles must take the largest possible step in space exploration toward the ultimate goal of detailed planetary exploration and full knowledge and control of our space environment.

What, then, must be the steps in our space program to achieve the promise that the 1970s hold? With the Thor-Agena and the Atlas-Agena vehicles just around the corner and ready to carry a ton or two, we can press the first few space measures vigorously.

1. The exploration of space itself must be pressed with full scientific instrumentation to delineate the fields and particles in space and to define unambiguously the conditions for safe manned flight in the future. The Thor-Agena and Atlas-Agena vehicles should be adequate and must be available in sufficient numbers to do the job.

2. Our scientific leadership must be retained by sound, advanced, and ever-improving experiments on space astronomy, meteorology from space, and a whole variety of studies about our earth that our space vehicles and space technology now permit. Aside from its scientific importance, this program permits training of large numbers of scientists in participation in space problems.

3. The commercial applications of space to communications and navigation are justified on high priority to provide to man the early dollar returns and daily benefits of our space program.

4. More sophisticated methods of guidance and control must be introduced and proved so we can put our payloads where we want them. In particular, a high measure of stability control must be made available.

5. We must concentrate on providing and space-proving better electrical power supplies for space applications using both solar cells and improved thermoelectric devices on one hand, and small nuclear reactors on the other, with a view to having available power levels ranging from a few hundred watts to perhaps eventually a hundred or a thousand kilowatts.

6. We can program the first, unmanned, "soft" landing on the moon to land scientific apparatus to telemeter back the primitive findings about the surface, atmosphere, and interior.

Our National Aeronautics and Space Administration has already programmed most, but not quite all, of these preliminary steps! They are all imperative, and a sufficient number of vehicles, with backup in case of failure, should be financed in time to ensure the substantial completion of this segment of our space program by 1965. Toward the end of this interval, the present deficiencies in our foundation program of space science will begin to become apparent.

By 1965, the Saturn and the Nova with their 10 to 20 ton payloads should be available. These will permit unmanned landings on the moon and perhaps return of samples in the 1965 to 1970 period. Certainly, this interval can see large unmanned astronomical observatories in space. Toward 1970 man can contemplate his first voyages out into space, perhaps orbiting the moon at close range, and even perhaps orbiting a planet or two at safe distance to obtain firsthand data for future landings. Certainly the sun can be approached by unmanned vehicles to a safe distance for critical scientific studies with complex apparatus. The construction of a space station might be possible toward the end of this era, though many scientists now feel that the moon might be the best space station to use, for the moon provides an adequate supply of structural material for shielding from X rays and powerful particle radiation.. Whether eventually, with a big nuclear reactor to supply almost unlimited power, the moon can be made to supply food, water, and air in limited quantities to permit a reasonable facility awaits the first detailed studies of its surface with manned vehicles in the 1970s.

We have already spoken of the first planetary attempts in the 1970 to 1980 era. But whether this program is achieved depends on our support of the research and the technology to effectuate it. Quite aside from present plans to instrument and operate our space vehicles over the next 5 years lies the tremendous need for scientific research to make sense of the program beyond the 1965s and especially in the 1970 era.

Let me raise some questions.

1. What are man's capabilities for doing anything useful under the extreme stresses of space travel? Those of you who would be space astronauts should read Alfred Lansing's "Endurance," the magnificent tale of Sir Ernest Shackleton and his men who, shipwrecked, made

their way 2,000 miles over the frozen and open seas of the Antarctic to safety. Yet their cold was much less than the cold of space, their weight limitations were not so severe, they had natural supplies of food and water and air. There is an immense job to be done so that man can just survive away from the earth, with perhaps a bare margin to do some primitive but significant scientific work, and to observe and record what he sees rationally in order that the next party can benefit from his experiences. The whole problem of ecology in space, shielding, movement, living processes, travel on the moon or planets, and scientific training and manipulation must be studied intimately, and space hardware designed accordingly.

2. How can data best be returned with the inevitable power and equipment limitations? At the least, some analysis of the data must be done by machine and computer; at best, we should like to have to transmit only the scientific paper describing the conclusions of the space observer or experiment. Somewhere between these extremes we must develop means of data analysis and transmission from space to maximize the information received from a given payload.

3. Our first landings must not contaminate the planets, and they must provide precise data on microorganisms or other life encountered. Earthly contamination may spread quickly to hide or destroy the very greatest scientific jewels of planetary discovery—those organisms that have evolved on the planets quite independently of life on the earth. Therefore, all earthly contamination must be avoided by the use of skillfully planned experimental procedures and well-tested equipment until rather advanced biological analysis is completed. Very advanced forms of biological analysis will therefore be desirable from unmanned landings *before* the first manned landings. But this requires methods and procedures of science that as yet await development.

4. The problems of back contamination must be considered. Our own terrestrial biology has produced our oxygen atmosphere, for there was no oxygen in the atmosphere of the earth before biology decomposed sea water to provide it. There appears to be no oxygen on Mars, but was there never any? Or did it disappear because of evolution of an organism that depleted that oxygen by fixing it into solids? While the danger seems very small of bringing back to earth an organism that would lead to fatal epidemics, or might even destroy our food or air, it is nevertheless finite. Therefore, the first manned, or hopefully even

better, unmanned landings, must provide us with sufficient evidence that no danger of back contamination exists before samples, or even men, are allowed to return to earth.

The preparation for lunar and planetary studies is immense and must be far advanced before the necessary space laboratories to do the job can be designed. Moreover, solutions to these problems will inevitably control the international order of space flights as we approach the 1970s. We can expect that there will be complex problems of international regulation.

I could go on to outline a whole series of scientific projects that must be done on earth to make our long-range program in space both sensible and economical. *The main problems of space are tied to no particular flight at this time.* Without a strong program of scientific research, these basic problems will never be properly solved—we will destroy the most precious scientific evidence the planets can give us and conceivably, though not probably, destroy ourselves.

What is needed is a scientific and laboratory effort involving representatives of every element of our best scientific groups. The scale of research on the earth precedent to planetary flights should be comparable to that of the Manhattan Project and the Atomic Energy Commission. As a part of this effort, every major university must participate, and we must have some central scientific research laboratories to concentrate their major efforts on specific elements of space science as distinguished from science in specific payloads. While in-house government efforts are indispensable they are not enough, for you cannot recruit nearly all the scientific leaders that we need with the civil service forms. Industrial laboratories will play an important role, but we must also recruit leaders in the academic atmosphere of the university and the national laboratory. With this diversity to support a really broad effort in space science, the job can be done with distinction. But I reiterate that our pure space-science effort on the ground must be broadened quickly, and the cost before the end of the decade must grow to the order of ½ billion dollars annually to make the other 2½ billion effective. This broad program to give us a real foundation under intelligent space flights must start now!

In closing, you will ask why I have not mentioned nuclear power for rockets. Clearly this has much promise, but it is a difficult and long range project. Certainly, if we have a cessation of nuclear tests, we should hold Los Alamos together with its skills and enthusiasms against a possible violation of agreement. What better way to do it than to continue Project Rover and related projects for nuclear propulsion of space vehicles. Since we must have Los Alamos anyway and must give it a real job to keep its scientists and their enthusiasms, an even larger effort would involve minimal costs. But we should not expect results from nuclear propulsion too quickly—perhaps in the 1980 to 2000 era. When nuclear space propulsion finally comes, it will bring a new era to space exploration when landings on the planets will become routine.

2

Impact of the Present World Situation on the Development of Peaceful Uses of Space

JAMES H. DOOLITTLE

CHAIRMAN OF THE BOARD

SPACE TECHNOLOGY LABORATORIES, INC.

A.B., University of California, 1922; M.S., D.Sci., Massachusetts Institute of Technology, 1924–1925. U.S. Army Air Corps, 1917–1930; Lt. Gen., 1944. Led the first flight on Tokyo, 1942; Commanding General, Twelfth Air Force in North Africa, 1942; Commanding General, Fifteenth Air Force in Italy, 1943; Commanding General, Eighth Air Force in England, 1944; Commanding General, Eighth Air Force in Okinawa, 1943. Vice President and Director, Shell Oil Company—Vice President 1946–1958; Director, 1946 to date. Effective January 1, 1959, Board Chairman, Space Technology Laboratories, Inc. Chairman, Air Force Scientific Advisory Board, 1955–1958; Member, President's Board of Consultants on Foreign Intelligence, 1955–1960; Chairman, National Advisory Committee for Aeronautics, 1956–1958; Member, Advisory Board National Air Museum, Smithsonian Institution, 1956 to date; Member, Defense Science Board, 1957–1958; Member, President's Science Advisory Committee, 1957–1958; Member of National Aeronautics and Space Council, 1958. Holds the Congressional Medal of Honor and numerous other decorations from the U.S. and foreign governments. Numerous honorary degrees. Member, The President's Engineering Advisory Council, University of California. Life Member, Massachusetts Institute of Technology Corporation.

THE EVENTS of the past few years, highlighted by the first Russian Sputnik, have brought a bewildering torrent of technological milestones—I have to be careful not to say millstones—in this infant space age. We frequently find it difficult to revise and to update our ideas as fast as today's developments demand. These technological advancements pose so many new and heretofore undreamed of possibilities that it is almost impossible for the layman—much less the professional aerospace man or his colleagues in the various scientific disciplines—to find time for pause and reflection on the new concepts, objectives, and values that have been thrust upon us.

From a purely semantic viewpoint, the terms "cold war" and "space race" bring to mind the condition of tension and conflict that unfortunately exists today in the world. But let us—if you will—look upon at least one positive, useful aspect of the cold war and the race for supremacy in outer space. If it were not for the present critical world situation, there would not be the incentive for the intensive development of the space potential that there is today. Space has become big business.

Illustrating this, the United States is budgeting more than 40 billion dollars this year (1959–1960) for defense. Included in this amount, I understand, are Department of Defense plans to obligate more than 5 billion dollars for the various missile and space systems. For the next fiscal year, the Air Force is calling for more than 2 billion dollars for expenditures on its ballistic missile programs, and the Navy has programmed almost 1 billion for its Polaris missile program. Moreover, the National Aeronautics and Space Administration is asking for nearly 1 billion dollars for its projects, compared to its ½ billion dollar budget for the current fiscal year.

The present world situation has stimulated us to far greater effort not only in space technology but in all forms of scientific endeavor. Except for the cold war, we in America would undoubtedly have continued

18

our customary attitude of complacency. (As a matter of cold fact we are not yet, in any sense, exerting maximum effort.)

We Americans are optimistic in temperament. We do not really extend ourselves until a true emergency arises. Seldom in our history have we adequately planned ahead or taken necessary—and sometimes obvious—preventive steps. We have waited for crises such as the one currently facing us and then belatedly taken remedial action.

In time, I am certain, man's inherent, insatiable curiosity would have driven him to explore the unfathomable regions of outer space. But today's military requirements have made necessary the expenditures of enormous sums of money on space technology and equipment which otherwise would have been difficult to justify from a purely civil viewpoint.

The cold war and the space race—and the traditional concept of competition, a keystone to the American way of life—have proved powerful stimuli. Entirely aside from the stockpile of space technology and hardware that competitive American science and industry have developed in a few short years, the greatest stimulus of all to competition is the competition for survival. And that is precisely the type of competition in which we are presently engaged.

The question could be raised: Does the continuation of the cold war suggest a greater or lesser chance of the world's developing peaceful uses of space? This is a difficult question to answer. It requires serious consideration. Let us look at some of the determinants.

The military arm of our nation, in addition to missiles, has need for reconnaissance, surveillance, warning, and communications satellites; but as yet it is not entirely clear what is the military use for deep space— a use which could justify an all-out crash program calling for unlimited expenditures. At the present time, deep space launching costs are very high and the life of space payloads relatively limited, to mention only two of the obstacles. This does not mean that a decisive use may not be found in the future. I think it will.

On the other hand, the huge military expenditures in the ballistic missile program and the lesser expenditures for our military space program will continue to result in civilian applications for scientific and commercial purposes—all for peaceful pursuits.

As James Straubel, publisher of the *Air Force/Space Digest*, writes in a recent issue of his magazine:

The so-called "space race" with Russia—more of an Olympics than a race—is a scientific crusade for military, political, cultural, and economic objectives. To date in the electronic revolution, the marriage of human intelligence and mechanical brains has produced an offspring made for terror and destruction. This first-born—the big ballistic missile—is as yet the only consistent user of space.

What of the countless man-made objects that will use space in the months and years to come? Here we find the great challenge: To use space in the pursuit of goals that will benefit—rather than destroy—all of mankind.

I, for one, believe that this challenge can and must be met. This will require thought and effort on our part and cooperation from the rest of the world. Without this international dedication to peace—which is presently being discussed, but I am afraid not very sincerely by the other team—the space race might even fan the existing tension. This brings us to the other side of the coin concerning the cold war's effect upon the peaceful applications of space.

In a study prepared by Johns Hopkins University for the Senate Foreign Relations Committee it was pointedly stated:

> . . . foreseeable technological developments will sharpen existing tensions between the United States and the Soviet Union and provide fruitful sources of new ones.
>
> Certainly the competitive exploitation of space opens a new arena for the conduct of cold, limited, and even hot warfare. The past controversy over radio channels for the International Geophysical Year is a forerunner of these space disputes; other predictable controversies will involve the extension of national sovereignty and territorial boundaries into space.

In summary, then, I think it can fairly be said that for some time to come the rate of space technology development for peaceful purposes will be a direct function of the military exploitation of space. For, unless we decide to give prime national priority to, space exploration, only our critical national defense needs can, over the long haul, adequately fund the myriad possibilities that lie ahead of us in the research and development phases of both peaceful and war-deterrent space technology.

This poses an organizational question but I do not, at this time, think it desirable that we here become involved in the discussion of whether we should have one or more national space agencies, and if but one, whether it should be civil or military.

There is, of course, considerable precedent throughout history to illustrate how ambitious military programs have paved the way for later peaceful applications of new technologies and discoveries. As early as the fifteenth and sixteenth centuries there were those visionaries who said that new and unconquered land masses of large size could be reached by sailing west from Europe. Their predictions sounded just as pie-in-the-sky-like in that era as descriptions of interplanetary and interstellar space travel do to some people today.

Too, during the preparations for exploration by sea and during today's consideration of placing a manned capsule into orbit and of trips to the moon, this question was asked: But aren't the risks too dangerous for the participants?

Whether the adventurers were iron men aboard wooden ships or modern men in "G" suits aboard space ships, nobody could say before the feats were attempted exactly what would be the risks experienced and what would be the rewards found at the end of the line. Personal risk is today minimized by research, development, careful manufacture, detailed planning, selection, training, and indoctrination.

Throughout history, scientific advances and technological developments—including those sponsored by the military—have always, eventually if not immediately, had a peaceful and usually a commercial application. In the early days of aviation, military people—schooled in ground and sea tactics—could not fathom the enormous potential of the airplane for either military or commercial usage.

I do not personally remember but history recalls that many militarists scoffed in 1909 when they looked over the first military plane. Its $30,000 cost, tiny in relation to space vehicle costs today, was in those days a truly large figure. The ship could carry only a pilot and a passenger at a speed of 42½ miles per hour for one hour. The military plane, circa 1909, had a useful life expectancy of only about 30 flying hours and its per passenger mile cost, expressed in terms of today's dollars, was around $80.

"Airplanes," the military men said, "are impractically expensive. Their payloads are so small as to severely cripple their military usefulness." That was the professional military man speaking. Those investigating the possibilities of peaceful applications of the airplane said that commercial air transportation was "economically impractical."

Yet—less than 50 years later—the speed of the airplane has increased more than one order of magnitude and, even more importantly, its useful life has increased by three full orders of magnitude. Economically impractical? Hardly. Today's per passenger mile costs for a jet have been computed to be only a few cents. This, against $80 only five decades ago. And I confidently expect that history will repeat itself in this, the age of space.

The applications of military technology for peaceful use are many. To mention only a few: Radar, developed by the British for strictly military purposes during World War II, is now an integral part of every modern airliner and of every properly equipped ocean-going vessel. Radar also has many other applications. The jet fighter plane, which proved itself in Korea, and its other military counterparts led to today's fast, reliable, comfortable, efficient jet airliners.

The first flight to Hawaii was made in 1927 by two military men, Lieutenants Hegenberger and Maitland. The hazardous trip required 25 hours and 49 minutes. Today's jet airliner enables you to eat breakfast in Los Angeles, fly to Hawaii for a swim in Waikiki surf, and return to Los Angeles in time for dinner. We could point out that almost every improvement of military aircraft and equipment has had a direct or indirect application to commercial aircraft.

And what of our modern ballistic missiles? The ICBM Atlas, Titan, and Minuteman; the IRBM Thor, Jupiter, and Polaris. Are they the forerunners of tomorrow's intercontinental, interplanetary, and interstellar travel? I believe they are.

Already such military boosters as the Atlas and Thor have made vital contributions to the success of peaceful explorations of space. It was an Atlas that launched "Project Score," the so-called talking satellite that broadcast President Eisenhower's Christmas message of peace to the world on December 18, 1958.

And the Thor has been used for a number of deep space probes as man endeavors to unlock the secrets of the universe. It was a Thor that boosted the now-famous Explorer VI "Paddlewheel" satellite, which gave us the first photo of the earth taken from outer space.

Space Technology Laboratories' later "Paddlewheel" satellite—Pioneer V—in 1960 established the tremendous record of broadcasting scientific data all the way out to 22,500,000 miles from the earth in its approach to the orbit of the planet Venus as man endeavored to learn more about interplanetary space. Experiments in this 94.8 pound space laboratory were conducted in such vital areas as micrometeorite detection, radiation, magnetic fields, and radio propagation.

I would like to say a little more about our national problems—and about military necessity—before going further into the assigned subject of this chapter.

The United States has long been committed to a policy of retaliatory deterrence, to be employed only in case we or our allies are attacked. As a nation we desire peace with honor. We are, however, committed to maintain our freedom and to protect our way of life. This can be accomplished—considering the threat we face—only through strength.

Problems of national security take on new dimensions in the space age. For one thing, deterrent power that really deters has to exist in a state of split-second readiness. It has to be safeguarded, through hardening, dispersal, or mobility, against a surprise first strike with nuclear weapons. Civilian indoctrination and protection, especially against fallout dangers, is vital to our ability to sustain and recover from a nuclear attack.

Our primary deterrent force at the present time is the Strategic Air Command. Its principal weapon is the long-range manned jet airplane. The life and the effectiveness of the jet bomber will be increased through the use of air-to-ground missiles. We will soon have, in quantity, a more modern weapon—the intercontinental ballistic missile.

In September, 1959, the Department of Defense announced to the American public that this Atlas ICBM had become operational. One of the most spectacular aspects of the Atlas story is the fact that operational capability was accomplished in slightly more than 5 years from the time that the ICBM "crash" program got under way.

Top national priority was given to the ICBM, and in 1954 the Air Force was made the agent of the Defense Department for this program. Time schedules were compressed so heroically by the Air Force—supported by American industry—that the Atlas became operational sooner than the most optimistic forecasts.

Not only that, but the capabilities of the Atlas—reliability, range, accuracy, and destructive power—also exceed the most sanguine hopes of the von Neumann Committee which formulated the original development plan. It is expected that the two other American ICBMs—the Titan and the Minuteman—will be equally outstanding as they develop operational capability, too.

Those of us in the military and scientific communities thought that the reliability of the early Atlas would be about 50 per cent. We were all happy to be proved wrong. As this is being written, the Atlas has successfully achieved all, or virtually all, programmed objectives in twenty-three of the past twenty-four launches.

Scientists—including many associated with Space Technology Laboratories, which had systems engineering and technical direction responsibilities for the Atlas—were initially striving for strategic requirements which at the very least could be described as difficult. There were many dissenters to the hypothesis that technology could produce an ICBM with a range of 5,500 nautical miles. Not only did American scientists meet that objective, but they substantially bested it. Today's Atlas has an actual range exceeding 7,000 nautical miles.

In 1954, when the speeded-up Atlas program really got under way, the hoped-for area of accuracy was estimated at about 5 miles. Again, American ingenuity and teamwork, laboring under pressure to maintain the peace, developed a guidance system with an accuracy, at extreme range, of less than 2 miles.

The exact warhead yield or destructive power of the Atlas missile is classified, but is tremendous.

The United States also has two intermediate range ballistic missiles—the Air Force Thor and the Army Jupiter—in its operational inventory as a force for peace. The Navy's Polaris missile program is making excellent progress and should become operational by 1961.

The Air Force Titan ICBM—2 years younger than the Atlas and therefore still subject to some growing pains—has been programmed to become part of America's deterrent force during the second half of 1961. And prior to mid-1963 we hope to have the Minuteman ICBM standing on its pads as an additional guardian of the peace.

We are already at work on second generation and thinking about third and even fourth generation space vehicles for a wide variety of space missions. Adaptations of them can serve the national security by providing reliable methods for inspection of arms control agreements. Such vehicles can be used to patrol peace as well as to deter war. For example: a surveillance satellite would help control the peace by affording the free world a new "summit" from which to observe the entire Planet Earth.

So far neither the United States nor Russia, as far as we know, has developed a positive defense against a ballistic missile fired in anger. This is one of our immediate needs in developing a more powerful deterrent to war. The most we can do now is to obtain early warning through the use of powerful ground radar to determine that a missile is on its way toward us and due to strike in about 15 minutes.

To obtain faster and more positive warning, the United States is developing a space-borne patrol which will greatly improve our ability to detect enemy missiles. These so-called very-early-warning satellites will give us an extra 15 minutes or more warning of enemy missiles being launched against the United States or one of its allies. Called Midas, and now being developed by the Air Force, the project uses sensitive infrared sensors which should be able to detect a missile shortly after it leaves its launching pad.

When you take into consideration that it may require 15 minutes to get one of our own missiles off the pad, the extra 15 minutes or more of warning time becomes a vitally important part of our deterrent power.

A third peace-patrolling space vehicle would be a military communications satellite. It would supplement and complement the work of the two previously mentioned. Communications satellites offer unusual possibilities as repeater sites. Due to the advantages of its location, a single station in orbit could replace hundreds of stations on earth and could span oceans as well.

Surveillance, warning, and communications satellites such as these could some day lead to a global warning network with direction and participation on an international basis. The free world's strategic attack system—alerted by space warning—would be ready for quick and decisive action against potential aggressors.

As Straubel has said:

> The next step could well be the pooling of the world's strategic attack forces into an international enforcement agency, geared to the global warning network, directed by the United States. With all the world alerted against aggressive action, the need for huge national attack forces would deteriorate, and voluntary reduction of armaments would be encouraged. In reducing the threat of surprise at-

tack, the program could eventually justify the curtailment of arms to minimum levels. This is the great utilitarian mission to be found in space.

Summarizing, then, we are making this multibillion dollar military investment in space for three basic reasons: (1) to deter aggression; (2) as decisive weapons in case of aggression against us; and (3) ballistic missiles and military satellites are the demonstrated forerunners of tomorrow's peaceful space vehicles—the transitional bridge between military and peaceful applications.

What, then, are some of the directly peaceful uses of space?

First and foremost, space is the medium where, through mutual consent, peace may best be monitored, enforced, and maintained. Moreover, several commercial applications immediately come to mind.

The first is a world-wide communication and television system. Three equally spaced satellites containing receiving and transmitting equipment and placed in a 24-hour equatorial orbit would supplement all or even replace some existing land lines and other communications means.

The satellites would be at a distance of about 19,300 nautical miles from earth and would appear to stand still in reference to a point on the earth. These three satellites would provide complete coverage for the earth, except the polar areas, and act as relay stations for worldwide communications.

Certainly the military communications satellite I mentioned earlier would offer vast potentialities for peaceful commercial usage. The Atlas talking satellite, Score, which I also mentioned before, while not a stationary 24-hour satellite, did demonstrate a technique useful for satellite relay communications.

And, touching again upon the peaceful application of pioneering efforts by military people, we are reminded that militarily supported studies of the nature of solar disturbances, of radiation, and of cosmic rays, and their effects on radio transmission, have helped greatly to improve the reliability of the Strategic Air Command's world-wide communications network. Problems that have long plagued radio communications are being solved. It takes no stretch of the imagination to appreciate the value of these studies to commercial communications and particularly to transport.

Another extension of the versatility of these satellites would be to the merchant marine fleets of the world. With a navigational satellite to guide them, ship captains could verify their positions regardless of overcast days and nights when they would otherwise be unable to obtain a "fix" from the sun or stars.

Knowledge of what lies beyond us in outer space already is proving to be a boon in the science of weather forecasting. Thanks to the many scientific space probes conducted by the free world, we have a far greater knowledge of the extent and nature of the atmosphere and the effect of external influences upon it. Our comprehension of the movements of polar air masses and the jet stream is already improving methods of weather forecasting and flight planning. In time, many lives and many millions of dollars may be saved by precise forecasts of hurricanes, typhoons, and other severe weather disturbances.

Meteorological satellites could observe and report world-wide cloud cover and greatly increase the accuracy of both long- and short-range weather forecasting. Also, it is not unlikely that we will some day be able to control the weather at will. These satellites will be one of the tools used.

A variety of satellites and space probes have been launched and dozens more are programmed. Answers to questions that have for decades been puzzling scientists are resulting from these space laboratories.

So far I have spoken primarily of these instrumented payloads. But man is of an adventurous nature. Merely sending packages of instrumentation into outer space will only serve to further whet his curiosity as to what he himself will find out there.

In its initial phases, space travel will be exploratory. The astronauts are pioneering in space exploration and in space transportation. Their experiments may well lead the way to economical intercontinental and transoceanic travel at hypersonic speeds. One day—and it may not be in the far distant future—we may be able to fly, or project ourselves, from Los Angeles to New York in half an hour and from Los Angeles to Paris in 1 hour.

It will, of course, be some time after such flights become technologically possible before they are economically feasible. (The prevailing westerly wind, obviously, will not affect space travel speed as it affects air travel speed today. Oddly enough, the same relative effect will result from the earth's easterly rotation.)

Later, I look for space travel away from the earth and eventually, perhaps, it may entail mass migrations. Certainly we can conceive of environments in the cosmos that are sufficiently similar to ours so that human life could exist comfortably.

It may even be possible to find or develop superior environments. Initially, we will study the environments of and then travel to the moon and the nearer planets in our solar system.

With the present and anticipated explosive rate of population increase on earth, we may be forced to people another world. After relatively near space is as common a visiting place as San Francisco is to Southern California earthlings today, then perhaps, in years to come, interplanetary or even interstellar travel may be no more difficult or hazardous than our pioneer ancestors found when they built the West.

I realize fully that any prediction of distant "futures" in science and engineering is difficult and uncertain. The extrapolation of missile and space technology is no exception. In some measure, this difficulty arises from attempting to separate those future possibilities which depend only upon continued engineering development of known scientific principles from those possibilities which require really new scientific discoveries. The basic determinant in the rate of development of any new project is the national priority assigned it; that is, the amount of our resources—mental, physical, material, and financial—devoted to it.

Obviously, the projects I have mentioned will be costly. However, we must realize that the mere addition of "X" number of dollars to our military budget does not, in itself, guarantee that an adequate deterrent to war will result. Nor is the current state of the art of space technology such that our civil budgeteers can assure that any given expenditure of money and effort will permit us to win a peaceful space race.

The real requirement is the will of the American public to compete, to sacrifice, and to provide a ballistic missile and space program for peace that will be second to none in the world.

May I call your attention to a striking fact. The Soviet economy is, in effect, a space economy now, while ours is largely a consumer economy. They are concentrating on the space race as much of their scientific, technological, and economic resources as they think necessary to win it.

They are doing this despite the fact that their gross national product is only one-half of ours—although some parts are increasing much more rapidly than our own. Moreover, their expenditures for the military establishment equal or surpass those of the United States. The Soviets intend to win the space race and eventually the economic and ideological race.

In 1959 I had occasion to visit Russia as a United States National Aeronautic Association delegate to the annual convention of the Fédération Aeronautique Internationale held in Moscow. We spent 8 days in Moscow and 9 days cruising around.

I was especially impressed with a billboard in the Crimea. This display succinctly outlined the Soviet economic objective: "By 1965 the Communist Bloc will account for more than half of the world's production." In other words, they intend to produce more than all the rest of the world put together. To stimulate their people to greater effort they promise them that: "Within ten years the Soviet citizen will be the most contented in the world." That is to say, he will have a higher standard of living than the American.

To counter this, we in America must produce more. We must employ improved methods. We must work harder.

We must, moreover, build up our stockpile of basic knowledge. This and the power and vitality of our economy are our most important national resources. Our most valuable national asset is the brain power required to build this stockpile. We must concentrate on locating it, developing it, directing it.

Unfortunately, we have so far failed in this. The low esteem in which education has been held for years is evident from the dollar value we have attached to its support. The consequence is a serious shortage of classrooms, of teachers, and of highly trained people—particularly in science and technology, in research and development.

The average teacher's salary of $5,000 per year falls some 60 per cent below the average income prevailing in seventeen other comparable professions. The wonder is that we face only a teacher shortage rather than a teacher famine. This can only be explained by the motivation and dedication of our teachers.

The basic difference between the Soviet system of education and our own is that they better understand the importance of education. The Soviet student is more disciplined. And he studies harder. Just look at their curriculum: From the first through the tenth (now eleventh) grades—precollege—the Soviet student takes mathematics every year and is well into differential and integral calculus by the time he graduates from secondary school. He studies physics 5 years, biology for another 5, and chemistry for 4.

Moreover, the Russian students' economic incentive for intellectual excellence is great. An outstanding professor of science, who is a member of the Soviet National Academy of Science, is reputed to get fifty times the remuneration of a day laborer.

Let us, on the other hand, examine where we stand in refining our most vital resource. Within 3 months from now, industry will require some 50,000 engineering graduates. Tabulations show that only 39,000 will graduate. The remaining 11,000 cannot be recruited. They cannot be recovered. They will simply not exist. This is only one indication of an even more serious problem: Student interest in engineering is lessening. Engineering enrollments are dropping.

We must exert sound leadership in this area if we are to be even remotely successful. We must find and encourage the "unusual"—the superior—student. Most important, we, as a nation, must appreciate and reward intellectual excellence. More scholarships and more fellowships will help.

A more intensive and skillfully directed program of public education—on education itself—must be started at once. At the high school level, students must be made aware of the opportunities—challenging and rewarding from the point of view of both financial and personal satisfaction—that exist within the scientific and engineering communities.

Let's take the lead in supporting a renaissance in learning. Let's insist that the scholar get at least the accolades accorded the athlete. Let's call a halt to the march toward mediocrity. Let's make our school courses sufficiently difficult that the average student must study hard—so that even the gifted does not find it so easy that laziness is encouraged.

To reiterate: We must find, encourage, develop, and utilize genius. Our success in every major endeavor, whether in social and political life, or in technological advance, depends largely upon what we do *now* to elevate American education.

In a technological race such as we now are in with Russia, there is of course a very great propaganda value to successful accomplishments in space. Weakly committed and uncommitted nations are carefully watching and comparing Russia's achievements with our own—although too often their grounds for comparison are predicated upon the volume of the voice of the Soviet propaganda machine.

It would be absurd to say that Russia expects to win converts to communism on the Moon or Mars. They are using their space "firsts" in an effort to win converts here on earth—in the Middle East, Asia, Africa, Europe, and the Americas.

Space is a propaganda field in which Russia chose to joust as part of their cold war effort. They started an aggressive endeavor in 1946 and it was not until 1954—8 vitally important years later—that we started our own high-priority program.

There is an important difference in their objective and ours in this matter. When we probe space we insist—and rightly so, I believe—on getting for our investment a high yield in scientific information. And we weigh such investments against alternatives that might produce bigger dividends in national security and national welfare. By contrast, the Soviets measure the value of their space ventures, regardless of any other results, primarily in terms of waging and winning the cold war.

We would be less than realistic if we overlooked the fact that we have also, as a nation, achieved a number of scientific firsts during these initial years of the space age. In some ways we have permitted ourselves to be thrown off balance by the scientific announcements coming from within the Soviet Union.

There are thousands of applications of space technology. We can't for a moment hope to read Soviet minds and attempt to determine which they will next choose to exploit. Moreover, we must guard against jumping to the conclusion that the latest Russian "first" has actual scientific or utilitarian goals when, in point of fact, their feats may have been designed solely for global propaganda purposes.

Viewed in retrospect, our record of scientific achievements in such a comparatively short period of time is excellent. The United States has commanded and communicated with an interplanetary body—Pioneer V—far beyond the range attained by any other nation.

We precisely guided into orbit a very heavy satellite containing complex communication relay equipment—the so-called talking satellite, Score. We have fired two animals farther out into space—some 1,220 nautical miles above the earth's surface. Telemetered data indicated that these animals—two white mice—entered our atmosphere alive, and survived impact.

Certainly, we must not merely follow Russia, and ourselves attempt to improve upon their feats. This is a defensive attitude. And in war or peace, the side which operates from a purely defensive standpoint is almost always at a distinct disadvantage.

Early in 1960 the director of the United States Information Agency told Congress that we are losing prestige throughout the world because of Soviet successes in space. There is already a tendency in world opinion, said George Allen, to view the Soviet Union as pre-eminent in all fields of science and technology because of its space feats. Furthermore, he added, these successes have created a "cockiness" among Soviet officials that endangers world peace.

The point is well taken. The present world situation demands that we not only compete with Russia in space but in all other forms of science and technology as well. To win the world for freedom and permanent peace takes more than weaponry. It calls for the application of science and technology to human needs; to making the deserts bloom with desalinized sea water; to converting solar energy to power; to unlocking the secrets of the living cell to achieve a longer, healthier life span; and eventually through the life sciences, particularly social science and behavioral science, to developing better understanding and cooperation between peoples.

If we are to mold our prestige in the Middle Eastern countries, then sending up a very heavy satellite into orbit around the earth will contribute a great deal. But so would our science and technology devoted to the development of a cheap source of fresh water for these peoples.

Obtaining scientific data on the Van Allen radiation belts around the earth is important, too. But so is obtaining better data on the nature of the earth's crust, its oceans, and its atmosphere. Or better data on the effects of human fatigue.

While we continue to devote our best scientific efforts to space technology, we must not—as we develop this concurrency of competition along all fronts—lose sight of the distinct possibility that Russia is hoping we will concentrate all our military attention upon space. There are other deterrent-to-war approaches which demand attention.

While I am of course profoundly concerned that we devote our whole-hearted efforts to gaining a technical advantage in space weapons, I am also positive that we must not permit their technology to steal in our back door, as it were, with some aggressive weapon equally or even more effective. For example, it is not inconsistent with their philosophical outlook that part, at least, of Russia's stated willingness to abandon nuclear weapons is compensated for by her development and possession of chemical and biological warfare weapons of a highly lethal and selective nature.

Space—first near and then far—will be used for military purposes, scientific discovery, exploration, and transportation.

Our primary purpose—world conditions being what they are—must remain that of maintaining our posture of positive military security as a forceful deterrent to war. Military expenditures in the ballistic missile and space programs will continue to be large. The military programs will have applications to, and a profound influence and stimulation upon what is available for the pursuit of, purely peaceful applications of space. Historical precedent clearly demonstrates that technological advancement for peaceful applications is a direct follow-on to military pioneering—the only area where ample funding at the start is generally available.

Only if there is effective, universal disarmament or a greatly increased national interest in space exploration can we anticipate the rapid development of peaceful space technology completely free from military support.

We must not let the Soviets' space achievements send us into paroxysms of imitation. We must seize the initiative and work harder, not only on our space projects but in all other areas of technology and science.

We must demonstrate greater regard for the "eggheads" among us and develop a greater appreciation of and stimulus for education—the wellspring of our nation's primary asset: brain power.

It is obvious that the Soviets can outdo us in selected fields by investing more of their resources in a particular activity than we are willing to invest. There is, however, not the slightest doubt in my mind that by the very nature of our people and the power and vitality of our economy we can still outdo them in *any* field of endeavor if we choose to concentrate enough of our human and material resources on that objective.

Our goal is peace on earth. We hope that space technology, although necessarily employed at present to deter war, may one day be the medium through which nations will, by mutual consent, establish a lasting peace. To this end we must dedicate ourselves.

3

Communications in the Space Age

FREDERICK R. KAPPEL

PRESIDENT

AMERICAN TELEPHONE AND TELEGRAPH COMPANY

Mr. Kappel is a native of Albert Lea, Minnesota, and a graduate of the University of Minnesota, where he received his B.S. in Engineering in 1924. In the same year he began work with the Northwestern Bell Telephone Company, first in plant construction and maintenance, and soon in engineering. Rising through the ranks, he became vice president of Northwestern Bell, in charge of operations, in 1942. In 1949 he was elected to vice president of A. T. & T., where he first headed the company's Long Lines department, and later the headquarters operating and engineering staff. In 1954 Mr. Kappel became president of the Western Electric Company, and in September of 1956 he was elected president of A. T. & T.

Mr. Kappel received the Outstanding Achievement Award of the University of Minnesota in 1954, and has been awarded honorary degrees by Lehigh University, Knox College, Union University, and Ohio Wesleyan University. He is a Fellow of the American Institute of Electrical Engineers and a member of Eta Kappa Nu, honorary engineering society. In 1959 the Economic Club of New York awarded him its gold medal for excellence of management. His directorships and trusteeships include the American Telephone and Telegraph Company, the Chase Manhattan Bank in New York, the Presbyterian Hospital at the Columbia-Presbyterian Medical Center in New York, the Committee for Economic Development, the National Safety Council, and The Tax Foundation.

ON AUGUST 12, 1960, a scientist at the Bell Laboratories in Holmdel, New Jersey, talked over the telephone with a scientist of the Jet Propulsion Laboratory at Goldstone, California. On that day, the people of the United States were making, all told, perhaps 275 million telephone calls. The conversation between the scientists, however, was different from all the others, and the very first of its kind. For they were talking by way of satellite.

In Holmdel the talking circuit was connected to a radio transmitter dish 60 feet in diameter. This sent out the voice in the form of radio energy. Our scientists beamed the transmission to a huge, aluminized plastic balloon satellite orbiting 1,000 miles above the earth. The shiny sphere reflected the radio signals, and some of the reflected energy was received at Goldstone. At the same time, the voice of the speaker at Goldstone was reflected from the satellite to a radio receiver at Holmdel. Thus there could be a complete conversation for a few minutes at a time —whenever the satellite was above the horizon in both places.

This balloon satellite was called Echo I and was sponsored by the National Aeronautics and Space Administration, the civilian space agency.

I shall describe the Echo I experiment in more detail, but first I would like to explain briefly why we in the Bell System have been putting our effort and money into satellite communications research.

Twenty years ago, in 1940, there were about 2 and one-quarter billion people in the world, of whom 132 million were in the United States. At that time also, the world had 42 and one-half million telephones, of which 21 million, or about half, were in this country.

Today the world's population is about 2.9 billion and our own population is close to 180 million. The total number of telephones is 133 million, and 71 million of them, or more than half, are in the U.S.A.

So, while world population has increased about 30 per cent, the telephones have more than tripled. And there is no doubt that in the years ahead all these numbers will continue to increase rapidly.

To put the prospect into a single sentence, as more and more people all over the globe have more and more telephones, they will inevitably want to make more and more calls across the oceans. So there will have to be a very large increase in the number of communication channels between continents.

Now let me draw a contrast between the number of long distance talking channels that go over *land* today and the number that go across *oceans*. I will give you some examples. Between the Los Angeles area and metropolitan New York there are more than 850. Between Los Angeles and metropolitan Chicago there are more than 600. Between the New York and Chicago areas there are more than 2,600. Our total land network provides some 250 thousand intercity voice channels, and we are adding more every day.

Across the oceans, however, the story is quite different. True enough we have made great progress in the last few years, so that across the Atlantic, for example, there are now twice as many talking channels as there were only 4 years ago. And in the past generation we have established sound, practical working relations with the communication agencies of nations all over the world. But the number of world-wide voiceways is still relatively small.

Now, the enormous expansion of the land network has come about largely by the development of what we call *broadband* communication systems. These provide hundreds and even thousands of talking channels at a clip. In the most modern coaxial cable system, for instance, we can provide 5,400 voice paths. Broadband microwave radio systems, using relay towers every 25 miles or so, have even greater capacity—in fact the very latest type can handle nearly 11,000 conversations. And in the laboratory we are developing so-called waveguide systems that will provide no less than 200,000 talking channels through a hollow pipe 2 inches in diameter.

The point I am coming to is this: Satellite communication systems will be broadband systems too. If we can make them work as we believe we can, they will help us increase the number of intercontinental voiceways many times over. We are not exploring the possibilities in order to get a dozen new circuits across an ocean, or even a dozen dozen. We start with the idea of getting perhaps 500 to 1,000 out of a single system, and hopefully, later on, a great many more than that. Moreover, if by using a number of satellites we can span one ocean, then using these same satellites plus a few more, we can span others as well. We can create several systems interconnecting different points, instead of just one. Another very interesting aspect is that it is precisely a broadband system—one that will transmit a wide range of radio frequencies—that is needed for world-wide television. The band of frequencies needed to handle a single television picture is hundreds of times wider than the band needed to carry a voice. On land our microwave radio systems and coaxial cables provide these broad bands. But to cross the oceans is another matter, for reasons I shall touch on later. Hence broadband transmission via satellite offers a most hopeful possibility.

Echo I travels about 16,000 miles an hour, circling the earth once every 118 minutes at a distance of 1,000 miles. Although the balloon is 100 feet in diameter and appears as a bright star at night, it is a mere speck in the sky—equivalent to a half dollar at a distance of 1 mile—and is not visible at all in full daylight. Our huge radio antennas must track it with extreme precision. In fact, the beam of radio waves directed at the satellite must be aimed accurately to a tenth of a degree. To deal with this problem, the participating laboratories and the space agency have built up a highly sophisticated, world-wide coordination of radar stations and computers.

This complex was tested in advance by sending voice signals from Holmdel to Goldstone several times by reflection from the moon. And radio energy transmitted from Holmdel was reflected—briefly—from the small Tiros "weather-eye" satellite.

The immediate purposes of the Echo I experiment were all in the area of basic research. They were five in number. First and most obvious, of course, was to demonstrate that speech *could* be transmitted. Second, we needed fundamental data about the propagation of radio waves through the upper atmosphere and out into space. A third purpose was to test the new components that were used in the communication system. Fourth was to test the balloon itself. Fifth and last, the modes of tracking had to be tested.

More broadly, the need for such research can be stated in these terms: Those who are faced with the task of putting space to practical peacetime use must separate the ideas that are possible from those that are impossible. And after that, we must separate from the merely possible those things that seem to be feasible. We must appreciate too that in electronic engineering we generally get what we pay for.

Bell Laboratories people fully understand this, but they know too that part of their job is the exploration of dreams. One of our scientists, John R. Pierce, got to thinking seriously about using satellites for communications back in 1954—three years before the first artificial satellite even went into orbit. He studied various ways of doing the job, and in a technical paper published in 1955 he presented the very first concrete suggestions for satellite communications. Rocket technology still had to catch up. So did some of the electronics technology. As a result, Bell Laboratories people went promptly to work on the electronics end, and since that time they have developed certain devices that we believe will have tremendous value in satellite communications—notably a new form of receiving antenna and new types of amplifiers.

At the same time, we are drawing on research done many years ago. Back in the early 1930s, for instance, when artificial earth satellites were unheard of, J. G. Chaffee, also of our Laboratories, jotted down in his journal a new idea that he thought might result in an extremely low noise radio receiver. It did just that, but the invention when applied to earth-bound communication systems was less economical than other methods that also produced excellent results. So for years Mr. Chaffee's invention, called a "demodulation feedback receiver," stayed on the shelf. Yet it turns out today to be the best thing available for receiving signals reflected from satellites. Past research often pays this sort of dividend.

The reason why the Chaffee invention is so important is that it enables us to reduce unwanted radio noise, in relation to the voice signals we want to receive, 100 times. Noise from the sky is one of the great problems we must overcome in producing good communication circuits by way of satellites. The problem is all the greater because the voice signals are so very weak. Signals reflected from the Echo I satellite, for instance, have an energy of about one-billionth of one-millionth of a watt.

Do you understand how feeble that is? I will try to dramatize it for you. If I were standing 10,000 miles out in space, and I had in my hand a 1-watt lamp—like a pocket flashlight—the amount of light reaching you from that lamp would be about equal to the radio energy reflected from the balloon in space.

So in picking up this weak, weak voice signal, we must have extraordinarily sensitive equipment. And in bringing the signal up, we must make every possible effort to keep noise down. There is more to this noise problem than I have stated here, and I shall say more about it later. For the moment, however, this is enough to show how useful it is to have a receiver circuit that enables us to reduce the ratio of noise to signal by a factor of 100.

What I have just said leads to another point for emphasis. I think it is now common experience for most people, when they talk across the country by long distance, to hear and be heard easily. What you telephone users want, and what we strive to give you, is really high quality transmission. To bring this about has taken years of steady effort, and requires a vast amount of complex apparatus, all tuned to work together in a unified, delicately balanced nationwide system.

In the early days of telephony, Mark Twain is said to have suggested that he would keep a daily record on the performance of his telephone. If he could manage to hear the roar of a cannon at the other end of the line, he would mark a cross on his calendar. If he could hear the crash of thunder, he would make two cross marks. And if the cannon and the thunder occurred simultaneously, and he could tell the one from the other, he would be really generous and make three marks.

By 1936, however—roughly a quarter of a century ago—long distance telephone transmission had greatly improved. It was about as though people were talking to each other across a distance of 35 feet. Today that distance has been reduced to about 6 feet, and we are still improving.

I put this accent on quality, and the need for quality and balance all through a coordinated system, for a very definite reason. I said earlier that we hope satellite communication systems may give us hundreds or thousands of voice channels across the oceans, and TV channels too. But quantity is not enough. We cannot make do with thousands of poor circuits. We have no interest in haywiring space. We are after volume *plus* the high quality required to maintain a high level of performance all through the integrated system.

You will understand, therefore, why the scientists at our Holmdel laboratory worked to achieve in this Echo I experiment a circuit every bit as good as the best long distance circuits we have today.

A few illustrations will help to describe and accent some of the things I have been talking about.

I said that on land, one way we obtain great numbers of circuits is by using what we call broadband microwave systems, employing very high radio frequencies. But these high-frequency signals generally travel in straight lines, and over great distances they will gradually leave the surface of the earth. Hence, as shown in Figure 1, we relay them in short segments around the earth's curvature, and at each relay station we also amplify the signals. Of course we cannot build relay towers out in the ocean deeps, nor do we have sources of electrical power out there to amplify the signals.

Satellite relay points, however, offer straight-line paths away from the earth and back again as shown in Figure 2. Thus they look attractive for transoceanic communication.

The launching of a balloon satellite enabled us to make our first experiments utilizing this principle. The plastic sphere (Figure 3) was inflated after it left the atmosphere to 100-foot size—as high as a 10-story building.

To carry the satellite into orbit, the space agency used the Thor Delta rocket, guided by the system developed by Bell Laboratories and Western Electric for the Titan missile. Once the balloon satellite was in orbit, Bell Laboratories scientists at Holmdel started their experi-

FIGURE 1. On land, one way of obtaining great numbers of circuits is by using broadband microwave systems, employing very high radio frequencies. But these high-frequency signals generally travel in straight lines, and hence must be relayed in short segments around the earth's curvature.

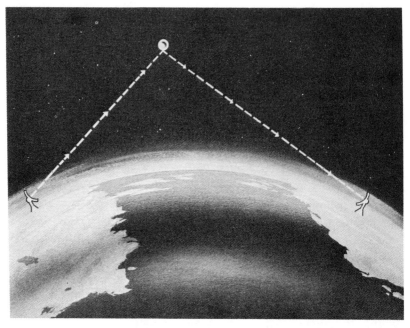

FIGURE 2. Satellite relay points offer straight-line paths away from the earth and back again.

ments. Near the foreground of Figure 4 you see the horn-shaped receiver antenna. This is the most sensitive voice radio receiver yet built. It employs a unique combination of the horn, a very sensitive amplifier, and also the Chaffee FM circuit that I mentioned earlier.

Figure 5 shows the present ocean telephone cables—indicated by solid lines—and also some of the cables proposed for the future. Early in this chapter I pointed out the disparity between the tremendous number of overland voice channels, and the relatively small number of overseas channels. But it is important also to emphasize that the ocean cables are just now starting to hit their stride. There is no question that developments now under way will enable us to increase the present number of telephone cable circuits several times over in the next 10 years, and it may well be that we can increase the number many times over.

The point is that to become commercially competitive a satellite system must not only match present and future earth-bound systems in terms of cost, it must also, as I have said, be of high quality—it must be dependable in providing uninterrupted service wherever it is needed—and we must use whatever techniques prove necessary to ensure the privacy of communications.

Earth-bound systems are constantly being improved. People first talked across the Atlantic by telephone in 1927, when the first commercial radiotelephone circuit was established. But only radio waves of lower frequencies, known as short waves, are suitable for long overseas jumps—we can use them because they bounce off layers of atmosphere and thus follow the earth's curvature. However, nature imposes very severe limitations. For one thing, these low-frequency radio channels are limited in number. Also, they are subject to interference by atmospheric conditions, and the 11-year sunspot cycle causes a lot of trouble.

Deep-sea telephone cables are free from these disturbances, and they provide a very high quality service—so good, in fact, that when we placed the first cable system in service between the United States and

FIGURE 3. Artist's conception of the plastic balloon satellite used in "Project Echo" experiments. It is 100 feet in diameter—as high as a 10-story building.

Figure 4. Bell Telephone Laboratories installation at Crawford Hill in Holmdel, N.J. Signals were bounced from here, via a man-made satellite, to Jet Propulsion Laboratory in Goldstone, Calif. Near the foreground is the unique horn-reflector antenna developed by Bell Laboratories as a receiver for the satellite communication system. The 60-foot dish-shaped transmitter is shown in the rear.

England in 1956, the number of conversations between the two countries nearly doubled in the first few months.

I mentioned earlier that coaxial cables on land, like microwave radio, provide very broad bands of frequencies. Hence they can handle thousands of conversations and also television. Ocean telephone cables are coaxial cables too; so it is natural to ask, "Why don't *they* carry thousands of conversations, plus TV, the same way the land cables do?"

A large part of the answer is this: Every such cable, on land or at sea, has to have electronic "repeaters" spaced at intervals along its length. These repeaters, as their name implies, repeatedly amplify the fading signals so that voices emerge at the other end strong and clear. On the high-capacity land cables, we use repeaters every few miles. But on a long distance ocean cable we have to space them much farther apart. You see, each one has to be supplied with power, and we can't get power from nearby. It has to come from the shore ends, and there is a limit to the amount of power we can provide over the length of the cable.

For such reasons, existing ocean cables have less capacity than cables on land. Today, for instance, we have two cable systems across the Atlantic. Each system uses two cables—one for each direction of transmission—and each has channels for about 80 conversations. But I call your attention to the fact that the capacity of ocean cables will grow markedly in the future. For instance, our next cable system across the Atlantic will handle about 230 conversations in a single cable—nearly three times as many as the existing twin-cable systems. At longer range, we are working toward cables that will use transistorized repeaters, rather than the electron-tube repeaters now employed. The logic of the problem is roughly this: To increase the capacity of the cables, it is necessary to use higher and higher frequencies. This in turn requires more closely spaced repeaters, and you will recall what I said about repeaters needing power. Transistors use much less power than electron tubes, but we must be sure that we can get high-frequency transistors that will be reliable enough to work for years at the bottom of the sea. When we are sure of this, then the capacity of ocean telephone cables will be due for another big jump.

I have spoken about the progress and problems of ocean cable development because, broadly speaking, the cables might be said to be the competition for satellite systems. Now let me get back to the satellites and speak further about some of the problems and alternatives they

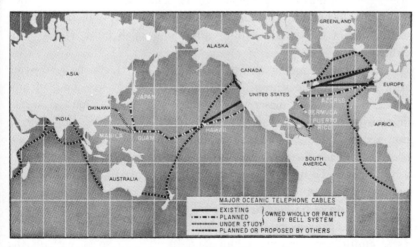

FIGURE 5. Present ocean telephone cables (solid lines) and some of the cable routes proposed for the future.

require us to consider. I'll begin with a word I have already used. That word is *noise*.

In radio, what you want to hear is the signal. What you don't want to hear is radio noise, because this makes it hard to hear the signal and may even drown it out. So what we are always after is a high signal-to-noise ratio—the most signal and the least noise.

Radio noise comes from various sources. You hear it as a background hum or hiss in your radio receiver. One of the main sources of noise is the receiver itself. But we have also found that an irreducible amount of noise comes from the sky—both from the atmosphere and from the emanations of bodies in the heavens. The fact that the stars emit this radio energy was discovered in the early 1930s by a Bell Laboratories scientist, Karl Jansky, who was then trying to pin down the source of noise on the newly installed overseas radio circuits. This discovery happily opened up the whole new field of radio astronomy. But in radio, noise from the heavens haunts us still, and especially in space communications. For though these emanations are slight, so are the signals reflected from satellites. Just keep in mind that billionth of a millionth of a watt!

Noise can also come from the area immediately surrounding the receiver; the unusual design of the horn antenna shown in Figure 4 is in order to keep this noise down to a very low level. But there is another essential unit in this whole receiver complex; this is the amplifier we employ to build up the faint, faint signal. Most amplifiers make lots of noise. How do we take care of this? How much noise will our amplifier make?

The really astounding answer is—almost none. I can give you this answer for only one reason: We have a completely new kind of amplifier. It is called a "maser." "Maser" is short for "Microwave Amplification by Stimulated Emission of Radiation." In 1957, following a proposal by Professor Nicolaas Bloembergen of Harvard, Bell Laboratories men made the first maser using solid materials. The heart of it is a ruby crystal. To make it work effectively, we have to cool it—chill it—way down to a minus 456 degrees Fahrenheit—close to absolute zero; and this we do with liquid helium.

The maser amplifier reduces by a hundred times the lowest previous noise emitted by a receiving amplifier. In all seriousness, without the maser we do not see how we could even hope for high-quality space communications.

Yet even with it, we have to do other things too. To bring this out for you, I shall again use a phrase I have used before. This phrase is "broad band."

The broader the band of frequencies, the more information we can transmit, other conditions remaining the same. To get our voice channels we cut the band into slices like a loaf of bread—one slice for each channel.

However, if the distant signal is very, very weak—and you know now how weak our signal is—we can also use *part* of the broad band to compensate for that weakness. We can manipulate the band to get a better signal-to-noise ratio. We not only can, we must. This is what the Chaffee FM receiver that I mentioned does. And of course there is a penalty. Each slice in our loaf has to be a bit thicker. From the total loaf, we get fewer slices.

In other words, for each good talking channel we hope to get, we need a lot of frequencies—many more frequencies per channel, for instance, than in an earth-bound microwave radio system. As I said, you have to pay for what you get—and in this instance the payment is in radio spectrum space. This has a very important bearing on our national policy and I shall say more about that later on.

Someone now may properly say, "You have talked about electronic repeaters on earth-bound communications systems; why not put a repeater on the satellite, so that you will get a stronger signal at the distant end?"

This is a good question. And the fact is that we *are* developing satellites that will carry repeaters. Such a satellite has already become known as an "active" satellite, whereas satellites which will merely act as reflectors are described as "passive." If we put an electronic repeater in a satellite to pick up and amplify signals from the ground, then on the ground we can use less powerful transmitters and smaller transmitting and receiving antennas. This of course makes for economy. However, in this case, the satellite must receive signals on one set of frequencies, and transmit them on a separate set—so in effect you are using up *two* slices for each channel.

Also, in designing any new radio communication system, we must keep it isolated, so to speak. It should be free from interference by other radio systems, and should not itself interfere with them. This latter problem tends to restrain us from making any tremendous increases in transmitter power. And not the least consideration is what all the costs will add up to, compared with alternative systems.

Weighing these factors, the engineers must also weigh various methods of using satellites. By now most of us are somewhat familiar with the basic principles of satellite physics. Satellites at lower altitudes complete their orbits faster than do those at higher altitudes. Naturally the lower ones would give us a more powerful radio signal. But the lower satellites are not visible simultaneously from any two points on earth for long periods. Echo I, for instance, at an altitude of 1,000 miles, is visible from both Holmdel and Goldstone for only 16 minutes during each orbit. Furthermore, low satellites are visible from a smaller area on earth at any one time.

This means that to maintain continuous communication, you need one satellite after another so that you will have at least one in sight all the time. Service to Europe, for instance, would require from 20 to 25 satellites orbiting a few thousand miles in space.

However, there are ways to make the laws of nature work for us. One way is to place a satellite in orbit at the precise distance required to make its period of orbit exactly 24 hours. If you placed a satellite at this distance—22,300 miles—so that it revolved about the equator from west to east, it would just keep pace with the earth's rotation and would appear to stay at the same point in the sky.

Only one such satellite would be needed to link any two points in the tropical and temperate regions of one-third of the globe. With three of them, as shown in Figure 6, we could reach all the world except the polar regions. However, a larger number would be needed to provide "spare" facilities in case of any failure and also to avoid wasteful use of microwave radio frequencies.

We may also find it possible to use the forces of nature to solve another problem. If antennas or specially shaped reflectors on satellites could be made to point always toward the earth, they could certainly increase the signal strength. However, positioning them in this way might use up power that could better be employed to amplify the signal.

One of "nature's own remedies" has been proposed, as shown in Figure 7. This is to cause the satellite to spin when it leaves the rocket.

Figure 6. A satellite in orbit at a distance of 22,300 miles, revolving about the equator from west to east, would just keep pace with the earth's rotation and would appear to stay at the same point in the sky. With three of them, all the world except the polar regions could be reached.

It would then stabilize like a gyroscope. Two antennas—one at each end of the spin axis—would always stay in the same direction in space. If this axis were parallel to the earth's surface, and at right angles to the direction of travel, the antennas would always be oriented properly.

Going back now to the 24-hour satellite, it would present us with some important difficulties. First of all, it takes extremely delicate rocket control to place a satellite in such a precise orbit. Then, the gravity forces of sun and moon, and other forces, would tend to pull it off station. So you would need some little jets or other station-keeping propulsion aboard, and methods of controlling them.

Also, the great distance—22,300 miles—is no help at all in receiving a strong voice signal. Such a distance seems much too far for a "passive" or reflecting satellite. An active repeater would be needed. And the satellite would have to be oriented so as to turn the same face always

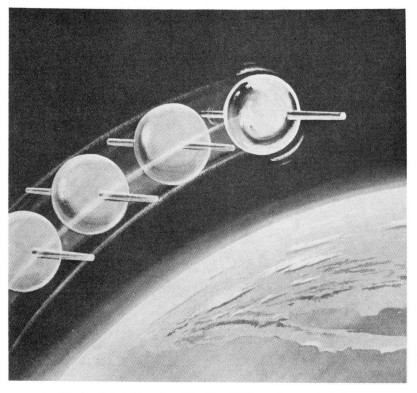

FIGURE 7. Satellite spin utilized to stabilize antenna orientation.

toward the earth. Furthermore the distance is far enough to impose a noticeable time lag. A two-way conversation would be delayed six-tenths of a second—long enough to be disturbing to the talkers. This would degrade the quality of telephone service, and it does not seem sensible to create a satellite system that would provide service inferior to that provided by existing ocean cables.

If satellites should carry repeaters, the size and cost of ground equipment would be greatly reduced. But the power sources and components of the repeaters would have to be very reliable and long-lasting. If anything fails, we can't make a service call out in space to fix it. Also, once you launch a satellite with a repeater built to handle a specific range of frequencies, and to operate in a particular way, you cannot change your mind later.

Another suggestion is for a satellite that would disperse thousands of short wires in a continuous band around the earth. These would reflect radio energy of the proper wavelengths, and there would be no need to track a satellite in order to transmit. Such a system might provide rough-and-ready narrow bands for military use. But it is believed to offer relatively few channels.

For active satellite repeaters, several methods of power supply have been proposed. One of the most reliable is a combination of storage batteries and solar generating cells. Solar cells, which were invented at Bell Laboratories, are semiconductor devices that use the sun's energy to produce electricity.

Solar cells furnish power to the small, narrow-band transmitter in the Vanguard satellite, which as this is being written is still sending out beep signals after 2½ years. Yet here too we need additional information. The belts of radiation in the sky can adversely affect solar cells unless the cells are protected. And we don't yet have adequate knowledge of the amounts and kinds of radiation present at different altitudes.

High-frequency electronic components with long life must be developed. Even the long-life components now used in submarine cables are not much help to satellite communications, because they are low-frequency components. Sky noise is much greater at the low frequencies, so it is better to use high ones. Also, antennas for low frequencies must be a good deal larger.

The larger the antenna, the more unmanageable it becomes for tracking a satellite—and the more it costs. The ground site at each end of a communication system associated with any but a 24-hour satellite will need four antennas—two transmitters and two receivers. One of each would be working while the others were making contact with the next satellite.

We should note too that complex auxiliary equipment, such as radar, is needed to make contact with the satellites and track them smoothly. Computers must keep tab on all orbits and furnish transmitting and receiving stations with schedules. At remote transmitting and receiving stations, power generators would be needed, fuel and other logistic support, and ground communication links to the continental telephone systems.

Now you know some of the principal factors involved. Suppose the problem were yours—what sort of system would you devise?

You might try to make some shrewd guesses. But before investing millions of dollars, you would want to investigate the possibilities further and accumulate as much knowledge as you could. That is what the Bell System has been doing.*

I have deliberately avoided building up the idea that the whole future of world communications is necessarily hitched to satellites. As I have said, another system—the undersea cable—is just hitting its stride. Yet I do not think it is visionary to say that the prospects are good that we shall succeed in using satellite communications. Satellites would give us an alternate means for doing all the things we do by cable today, and we can also hope that some day they will enable us to do even more. The transmission of live television, for instance, looks ultimately possible.

On the assumption that satellite communications *will* prove feasible for regular commercial use, I would like to touch on a few major questions that will have to be resolved in an orderly way.

These questions are both domestic and international. Some of the most crucial arise from the fact that to communicate via satellite it is necessary to use radio frequencies in huge quantities. The radio-frequency spectrum is of course a natural resource, and the frequencies that can be used for satellite communications are limited. The best area is from roughly 1,000 to 15,000 megacycles, with some possibility also of using the range from 15,000 to 20,000. In the extremely high frequency range, rain blocks off transmission. In the lower range, sky noise increases, and besides, big bundles of frequencies are already in use.

Obviously, satellite communication systems must not interfere with each other, and matters must also be arranged so that other uses of microwave radio—uses that have nothing to do with satellites—will not interfere with and overpower the faint signals from space. To define the uses of frequencies in various regions, we shall need the most thoughtfully considered international agreements.

Also, since the number of frequencies is limited, this will test the wisdom and foresight of each nation right now in charting domestic policies. In the United States, for example, there are at present no frequencies specified for space use in the frequency range most suitable

* Subsequent to Mr. Kappel's lecture, the American Telephone and Telegraph Company has proposed a system employing a number of active satellites orbiting a few thousand miles in space. The company has expressed confidence that such a system could be in full operation within a very few years, and offers the best means at this time for providing telephone service to the public by way of satellites.

for satellite communications—that is, the range above 890 megacycles. But there are any number of claimants who would like to use such frequencies for various purposes. Great care must be taken to make sure they are not wasted. It will be fruitless to create satellite communication systems, and at the same time create a shortage of the frequencies required.

Here is another question: What about the ownership and operation of satellite systems? People have several times asked me whether the Bell System, for instance, will some day own satellites. My answer is this:

I think we should share ownership of any space communication system we may use, just as we now share ownership of ocean cables with telephone companies and administrations overseas. This puts full responsibility for the service right where it belongs—that is, with the organizations that provide the service. As I said earlier, all parts of a communication system have to be designed and built and operated as units of the whole. Whether our overseas links go via space or across the ocean floor, the purpose is always to connect any telephone in this country with any telephone in another country. To do this well, we and our partners overseas must take the same kind of responsibility for the links *between* countries that we each take *within* countries. This is what is essential, and the particular type of system employed—cable or satellite—has no bearing on the question.

I don't know that the Bell System would go into the satellite-launching business. Today we ask contractors to build our buildings, and in the future we might ask other contractors to launch our satellites. That way we would still risk our own money—design and operate a system in harmony with all the rest of the telephone network—and be fully responsible and accountable for the service we provide.

Here I am looking ahead quite a bit. In the present stage of experiment and exploration, with space vehicles being launched for many different purposes, it is natural for the government to take the lead in putting up satellites. But when we come to providing communication service, I think we in the Bell System should take all the responsibility we can—under public regulation—for the job that is given us to do.

To be more emphatic, I am certain that this will be in the best interest of the United States. We need only look at the record to know that private enterprise has given this country the finest, the most complete, the most dependable communication system in the world. I strongly believe that private enterprise will also keep this country out in front in space communications, just as we have led the way on land and by sea. It was the Bell System, I may remind you, that pioneered world-wide telephony, first by radio and later through cables. We want and intend to be pioneers tomorrow too. We are ready, willing, and able to push ahead. All we need is the leeway, the freedom, to do our best.

I have discussed communications in the space age. I hope I have been able to convey my feeling that the future is loaded with exciting possibilities. But I must re-emphasize in conclusion that all world-wide communications in the space age will not be communications through space. Because the number of satellite systems we can create will be limited, a growing network of submarine cables will also be a necessity. And remember, as populations increase and there is need for more and more communications across the seas, there is room for an indefinite number of cables in the "inner space" of the ocean depths. Paradoxically, you see, there is more room for communications on earth than there is in space. In the long run, performance and economics will be the deciding factors. But a good estimate, I think, is that the two kinds of systems could very well complement each other; and certainly the existence of alternate systems would increase the over-all reliability of overseas communications.

There are challenges here for us all—for scientists and engineers, of course; for the management of the communications industry in every nation; for government leaders; and surely for educators, who must develop and train people for all these groups. Such is the age we are entering that I hope the general public, also, will gain a basic understanding of space technology and the broad problems and opportunities it presents.

I predict that by whatever means will serve the public most efficiently, we shall succeed in providing the requisites for a new era in world communications. And I leave with you the thought that in the space age, communications will play an ever-increasing part in the affairs and progress of mankind.

4

Space Technology
for World Navigation

VICE ADMIRAL JOHN T. HAYWARD

DEPUTY CHIEF OF NAVAL OPERATIONS (DEVELOPMENT)
UNITED STATES NAVY

B.S., U.S. Naval Academy, 1930. Graduate work in physics and chemistry at various schools including California Institute of Technology and Stanford University. Served through all commissioned ranks from Ensign to Vice Admiral during the period 1930 to present. From 1944 to 1948, Experimental Officer, Naval Ordnance Test Station, China Lake, where he worked on all phases of rocket development and study of destruction caused by the atom bomb. He received a Letter of Commendation from the Secretary of the Navy for extraordinary achievement in that field. From July to December, 1948, he was Director of Plans and Operations for the Armed Forces, Sandia Base, Albuquerque, New Mexico, concerned with the use of atomic weapons and integration of military requirements with the Los Alamos Scientific Laboratory. He became Special Assistant to the Director, Strategic Plans Division, Office of the Chief of Naval Operations in 1957 and in October, 1957, assumed the duties of Assistant Chief of Naval Operations (Research and Development). He holds numerous decorations including the Silver Star Medal, Legion of Merit, and Distinguished Flying Cross. In April, 1959, he reported as Deputy Chief of Naval Operations (Development).

S PECULATION ON the eventual influence of space on the lives of men is examined in many quarters these days. There are many legal, military, diplomatic, and scientific questions which arise as a result of expanding man's environment beyond the atmosphere of the Earth. Human thought and the general view of man's place in the scheme of things may well be profoundly influenced by man and his machines leaving the atmosphere of this planet. Just as the invention of the telescope led man to the discovery that he was not actually the center of the universe, astronautics and extraterrestrial exploration may lead to even more profound philosophical teachings.

The subject of this chapter is a limited but important portion of the space spectrum: namely, space technology for navigation around the world. I shall discuss specifically the state of navigation technology today, some of my personal speculations on future developments in this area, and the influence of space in these developments.

The History of Navigation

Since man's earliest journeys from his cave in the hills in search of food, he has found it necessary to navigate in order to survive. The guidance of a man or a vehicle from one place to another involves the two functions of *indication* and *control*. Navigation is the *indication* aspect of guidance. *The control* aspect of guidance involves the comparison of navigational data with programmed (stored) data, selection of the most appropriate action (decision-making), and execution of that action. Each one of us in driving an automobile to a particular destination, for example, carries out all the classical functions of navigation. If we know by heart the route from our home to the destination, the stored data is carried in the memory cells of our brain. If the route is strange, the stored data is probably utilized in the form of a map. Our actual position is determined en route by optical "piloting" methods; that is, by visual recognition of landmarks. A comparison is made of our actual position and our desired route or "programmed path"—if a difference exists at any time, a decision is required in order to nullify or erase the difference. Thus, by intermittently or continuously performing these classical steps which are inherent in all forms of navigation, each of us is able to arrive at our destination.

This example of a navigational system may appear to be an overcomplication of a rather simple situation. Yet, such analytical examination is required in order to establish a basis for engineering improvement of this or any other system. This example does contain all the elements of any navigational system, including those for ballistic missiles, seagoing vessels, and air transportation systems.

There are two important requirements for any navigational scheme:

1. The geographical area must be properly mapped so that the positions of all points are established.

2. The location of the vehicle in the area must be established.

Space offers an exciting new way to accomplish both of these objectives. Before discussing some of the ways in which space is important for world navigation, I should like to review briefly other navigational techniques common today.

The earliest form of navigation was "piloting." Piloting consists of observing visually, with or without the aid of optical instruments, the relative location of the navigator with respect to well-known landmarks. We move from room to room in our home, drive to work, steer a ship up a harbor channel, and land an airplane by piloting techniques. Another common form of ship and air navigation is by "dead reckoning," a corrupted abbreviation of "deduced reckoning." This might be described as an educated guess as to where you are, based on estimates of course, speed, winds, and sea currents. Dead reckoning alone is seldom adequate except for extremely short periods of time because of the many errors and uncertainties that creep into the system; position is generally established periodically by more positive methods in order to correct for dead-reckoning errors which accumulate with time.

In ancient times the range of sea travel was very limited, largely because of the mariner's difficulty in finding his way back home if he ventured too far out of sight of land. Sea voyages in most parts of the world at that time were confined practically to coasting in sight of land or to short runs between islands and headlands. Seafaring was somewhat less circumscribed in those parts of the world where the winds blow steadily

58 VICE ADMIRAL JOHN T. HAYWARD

in one direction at certain seasons of the year. For example, in the region of the Indian Ocean, monsoons blow from the northeast for about 6 months of the year and then shift to the southwest for the other 6 months. Man very early learned to take advantage of this aid to navigation in getting from place to place. It was because of this that the Arabians and other people living along the Indian Ocean were probably the first to attempt long voyages out of the sight of land. For this reason also, they were well ahead of their contemporaries in the Mediterranean and in the Atlantic in developing aids to navigation. The Arabs and not the Chinese, as was believed at one time, were probably the first to use a rudimentary compass for maintaining a desired course at sea. The first compass was simply a small round bar of iron, magnetized by a lodestone, floating on a reed or piece of wood in a bowl of water. Even the approximate date the mariner's compass first appeared is lost in the mists of antiquity. There is no authentic record that the compass as an instrument of navigation was known to ancient civilizations before Christ, or to the early Chinese. It is probable that this greatest of early advances in navigation did not appear much before the year A.D. 1000.

The greatest single break-through in sea travel—and surely one of the most significant steps forward in man's progress over the ages—was the discovery of celestial navigation. Man's ability to fix his position on Earth by measurements from the "fixed" stars made it possible for him to venture safely beyond the sight of land; he was no longer forced to rely on piloting as his only method of navigation. This achievement ranks with the harnessing of fire as an energy source, the invention of the wheel and lever, and Newton's three laws of mechanics as significant milestones in scientific progress. Now it was possible to spread civilization to all corners of the world.

The use of the heavenly bodies for guiding the traveler on land, particularly in desert areas, goes back to remotest antiquity. The practice was soon recognized as equally well adapted to guiding the traveler at sea. The first seaman in recorded literature who actually steered by the stars was Odysseus. While it is true that he was merely the mythical hero of an epic poem, we know that his exploits embody traditional methods of seafaring of an era older than Homer's (about 850 B.C.). When leaving Calypso's island, he

. . . spread his sail to catch the wind and used his seamanship to keep the boat straight with the steering-oar. There he sat and never closed his eyes in sleep, but kept them on the Pleiades, or watched the late-setting Bootes (Arcturus) and the Great Bear. . . . It was this constellation that the wise Goddess Calypso had told him to keep on his left hand as he made across the sea. So for seventeen days he sailed upon his course.

By keeping the Bear on his left, Odysseus was sailing east. Since the Pleiades and Arcturus differed by nearly 11 hours in the time at which they crossed the meridian, one or the other would always be visible in the night sky, and sometimes both. The Pleiades, so Pliny tells us, rose with the Sun 6 days before the Ides of May, so that by the time autumn came they were visible all night. Arcturus, which was first seen 40 days before the autumnal equinox, was high in the sky when the sailing season opened in the spring. In Homer's time this was a zenith star in the mid-Mediterranean Sea, and was termed "late-setting" or "slow-setting" because it only remained below the horizon for 6 hours out of the 24. Watching these familiar stars as they neared and crossed the meridian, Odysseus could steer due east until he picked up a familiar landmark.

Although this was but part of a story, it was a story told to seafaring listeners, and told by a poet who knew the ways of the sea and of sailors.

The ancient Greek philosophers taught that the Earth was round and that the location of any place on its surface could be described in terms of latitude and longitude. What the mariner lacked until comparatively recent times were the tools and data to determine his latitude and longitude at sea. Finding the approximate latitude of a place, particularly one north of the equator, was never very difficult, as it corresponds very closely to the angle of elevation of the North Star above the horizon. This fact was discovered even by primitive peoples. The early Hawaiians hit upon a contrivance, the Sacred Calabash, now preserved in the Bishop Museum, Honolulu, for determining when they had reached the latitude of the island of Hawaii in their return voyages from Tahiti. The Sacred Calabash has holes in its periphery near the top. With the Calabash held upright, observations were made on Polaris through one of the holes across the rim. When that star appeared just over the rim while sighting through the hole, the time had come to change course to the west because the latitude of Hawaii had been reached.

Many years after these voyages had ceased and the Sacred Calabash had become a museum piece, measurements were made of the gourd. It was found that the angle was 19°30' between the horizontal and a line passing through a hole tangent to the rim. This is exactly the latitude of Hawaii.

Over the years various instruments were invented to measure the altitude (angle of elevation) of heavenly bodies above the horizon, the earliest being the astrolabe described in ancient Greek literature. This and the cross-staff were the main instruments of seamen for measuring altitudes until they were superseded by Hadley's quadrant about the middle of the eighteenth century.

The determination of the other coordinate needed to fix the position of a ship at sea, namely its longitude, presented much greater difficulties to the early mariner than finding the latitude; in fact, it does to this day. The use of a timekeeper for this purpose was proposed by the Flemish astronomer Rainer Gemma-Frisius in a work of navigation published in Antwerp in 1530. He pointed out that inasmuch as the rotation of the Earth is absolutely uniform at the rate of 15° per hour, the only thing necessary to establish the longitude of a place is to know the difference in time at any moment between it and the reference place from which longitude is measured. It was known how to find the apparent or sun time at the ship's position when Frisius made this suggestion, but no reliable marine timekeeper existed for keeping track of the time of the reference place. In fact, no universally accepted reference meridian was recognized at this time. The meridian of Greenwich, England, was later adopted as the prime meridian from which longitude is measured east and west.

The great increase in seafaring brought about by the discovery of America stressed the importance of finding a convenient way of determining longitude at sea and kept the clock idea alive from the time it was first advanced. Toward the end of the sixteenth century, Philip II of Spain offered 1,000 crowns for an accurate ship's timepiece. Later Holland raised the offer to 100,000 florins. One hundred years passed, however, without the appearance of such a clock. In 1714 the British government offered a prize of 10,000 pounds (later raised to 20,000 pounds) for any means of determining a ship's longitude within 30 nautical miles at the end of a 6 weeks' voyage. John Harrison, a self-educated Yorkshire carpenter, won the prize in 1760 after a lifetime of effort, but his clock, while a splendid timekeeper, was too delicate, complicated, and costly for merchant-ship use. In 1765, Pierre Leroy of Paris improved on Harrison's clock and produced a marine time-keeper with a mechanism embodying in rudimentary form all the essential features of the modern chronometer. The determination of longitude at sea had now become possible by computations which were not too difficult and complicated to be made by mariners.

Progress in navigation during its early history was almost as dependent on improvements in chartmaking as it was on improvements in position finding. The Mercator system of drawing charts was developed during the latter part of the sixteenth century. Maps and charts using this form of projection began to come into general use about 1630.

The publication of a nautical almanac was started in 1767 by the Naval Observatory, Greenwich. This did much to place navigation on a scientific footing. The teaching of navigation and the publication of textbooks soon followed, the most notable example being Bowditch's *American Practical Navigator*, the first edition of which appeared in 1802. Rapid progress was made in the nineteenth century in the improvement of navigational instruments and in aids to navigation. These have been continued in the twentieth century. The most important were:

1. Magnetic compass correction procedures
2. Sonic depth finders
3. The gyrocompass
4. Speed measuring devices
5. Time signals to provide observatory time to ships
6. Special tables to shorten the mathematics of position finding
7. Improved sextants (including the bubble type)
8. Application of electronics to navigation in the form of radio, radar, and loran

As the above enumeration indicates, the twentieth century has led to quantum improvements in navigational techniques, largely due to the pressure of air-borne requirements for rapid position determination during all types of weather conditions, and largely made possible by advances in electronics. Radio beacons provide direction information from any unknown position to a given radio station. Simultaneous bearings on two or more stations provide position fixes. Radar is an active electronics method for measuring range and bearing from any point to known landmarks.

Loran provides another common method for fixing a vehicle's position. An airplane or ship using the new type Loran-C can measure its position within less than a quarter of a mile when operating as far as 1,000 miles away from the transmitters. Using ground-wave propagated signals, Loran-C provides high precision fixes out to 1,400 nautical miles during the day and to about 1,000 nautical miles at night. Operating from sky-wave signals, the system can be used out to a distance of 1,800 miles during the day or 2,300 miles at night with errors of the order of 2 to 3 miles. Loran-C is a low-frequency (100 kc) version of the older Loran-A and employs phase comparison methods for increased position accuracy.

Navigation information obtained at infrared frequencies has also been shown to be a feasible technique.

A significant feature common to all the navigation methods mentioned thus far is that all rely on electromagnetic radiations; namely, optical, radio, and radar. These methods suffer from propagation disturbances such as weather and ionized atmospheric layers, and many are limited to line-of-sight ranges. Radio and radar systems are also susceptible to man-made jamming.

Inertial methods have led to the most revolutionary changes in navigation during the last decade. An inertial navigation system consists basically of two or three accelerometers mounted on a gyrostabilized platform, and some form of computer. The stabilized platform may be either a physical platform or a mathematical (phantom) platform. Disturbances due to imperfection of the gyros and accelerometers, commonly called "drift," are primary sources of error in inertial systems.

Major advantages of using an inertial system are:

1. Navigational information is based entirely on measurements made from within the vehicle; no contact with the outside world is required.

2. Streamlined guidance and control system design is feasible since certain measurements required in the navigation function may be used for other control functions.

The automatic star tracker and the atomic clock are two other navigational advances of importance during the last decade. The star tracker continuously measures the angular parameters of a star or other celestial body. Systems of this type are now in operation and are able to obtain considerable accuracy in fixing positions.

Operation of the atomic clock depends on the frequency stability of certain chemical elements. Frequency can be measured with great accuracy—and frequency is another way of stating time. Certain elements, such as cesium, have an invariant nuclear resonance in a radio-frequency electromagnetic field. Time may be measured to an accuracy of one part in 1,000,000,000,000 with the atomic clock.

Thus, we have seen certain distinct advances in the long history of navigation up to the present time:

1. The fifteenth century saw the wide acceptance of celestial navigation, which made it possible for man to venture safely beyond the sight of land and permitted world-wide exploration by sea.

2. The electronics boom in the twentieth century made exotic new navigational techniques possible; namely, radio, radar, loran.

3. Large-scale advances in the art of building gyros and accelerometers after World War II, and the pioneering work of Schuler in Germany and Draper at Massachusetts Institute of Technology, led to inertial navigation systems. These became operationally useful in the mid-1950s.

Coincident with year-by-year improvement in the art of navigation has been a steady growth in world sea traffic. Figure 8 shows the tonnage of sea traffic in millions of metric tons in recent years. The total world shipping data for 1958 and 1959 are not yet available, hence are not included on the figure. The United States data are broken down into exports and imports; the figures show that U.S. imports and exports comprise about one-fourth of the total shipping in the world—a rather remarkable indication of our dependence on the seas around us.

There is one thing I want to stress about the figures on U.S. shipping given in Figure 8. These refer to total tonnage imported into the country and exported from the country, irrespective of the flag being flown by the carrying ship. Figure 9 shows the percentage of total imports and exports which are carried by vessels flying the American flag. This

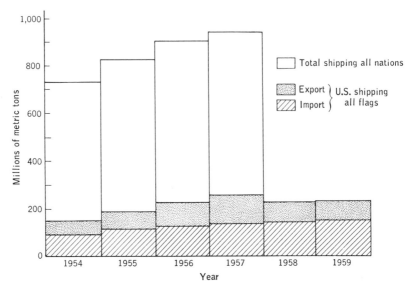

FIGURE 8. Tonnage of world sea traffic.

figure shows a very alarming trend: U.S. ships carried more than half of our cargo immediately after World War II, but the percentage has steadily and rapidly declined to less than 10 per cent in 1960.

History has shown that technological advances generally accelerate the economy and serve as healthy stimuli for trade. Technical advances in our shipping techniques, of which navigation with the aid of artificial earth satellites may be a part, are just the shot in the arm needed to reverse the slope on Figure 9.

I think an even greater indication of the need for an improvement in shipboard navigation is shown by the data given in Figure 10 of shipping accidents reported by the U.S. Coast Guard during the fiscal year 1959. Not all of these accidents were caused by navigation errors, but a very large percentage of them resulted directly from this cause. All of the 503 groundings, for example, resulted from gross navigation errors—and many of the 434 collisions were attributable to the same cause. The loss of lives and the loss in man-hours and dollars as a result of shortcomings in our present navigation system is tremendous.

I submit to you that one of the greatest steps up the ladder in the evolution of the science of navigation is taking place in the 1960s as man puts his own celestial bodies around the Earth to be near-permanent

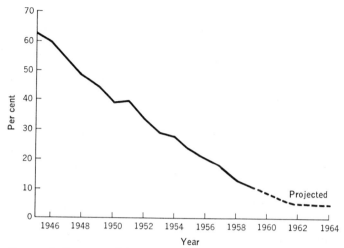

Figure 9. Per cent of United States exports and imports carried by United States ships.

landmarks for the guidance of vehicles over the surface of the Earth and through the air surrounding it.

Navigation has periodically taken advantage of advances in the disciplines and technologies of astronomy, geography, and physics. The development of rockets capable of launching artificial celestial objects into semipermanent orbits constitutes an advance which can be turned to the advantage of navigation. Systems utilizing artificial earth satellites can be visualized which are capable of the all-weather performance of radio navigation coupled with accuracies which equal or surpass optical celestial navigation. Other techniques of navigation such as dead reckoning, optical celestial sights during clear weather, and inertial

SHIPPING ACCIDENTS REPORTED TO U.S. COAST GUARD (1 July 1958 to 30 June 1959)	
Foundering, capsizing, sinking, etc.	171
Grounding	503
Collisions	434
Lives lost	200
Number of people injured	1,243

Figure 10

measurements can provide position information with adequate precision for limited periods of time. It is often necessary, however, to check independently and periodically the navigation data indicated by these systems.

Before proceeding to a more detailed discussion of satellite navigation systems, it might be well to review briefly some of the basic principles and terminology of celestial navigation. At any instant of time, every heavenly body—each star, planet, or satellite—is directly over some specific point on the surface of the Earth. This may be easily visualized if we imagine a straight line to be drawn from any such body to the center of the Earth. If an observer is standing at the point where this line pierces the surface of the Earth, he obviously will see the body vertically overhead (at the zenith) at that particular moment. This position directly beneath the celestial body is of course constantly moving across the surface of the Earth because of the relative movement between the Earth and the celestial body. Our knowledge of astronomy permits us to compute in advance, in most cases quite accurately, what the position on the Earth directly beneath any heavenly body will be for any specific instant of time in the future. Tables of these future positions for the more prominent celestial bodies are computed and

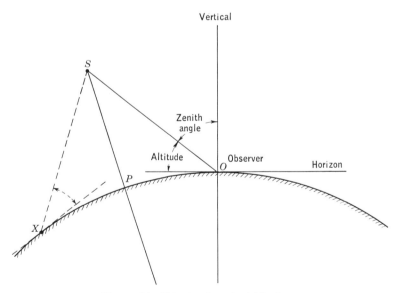

FIGURE 11. Altitude of a celestial body.

published for the use of navigators and are an essential part of celestial navigation.

A second concept is necessary in understanding the basic principles of celestial navigation. This is the angle of elevation of a celestial body above the horizon, or, in navigation terminology, its altitude. As shown in Figure 11, any observer O at any point on Earth looking at any celestial body S, sees that it forms a certain angle (altitude) with his horizon at a given moment. (Alternatively, the zenith angle may be measured; the altitude is equal to 90° minus the zenith angle.) If P is the point directly beneath the celestial body S, and if the distance PX equals the distance PO, then an observer at X would also see S making an equivalent angle with his horizon. If we now imagine ourselves sitting at S and looking at the Earth, we can see that there would be a whole circle of positions on the Earth's surface where each observer would see S forming the same angle with his horizon, as shown in Figure 12. (Another angle measured occasionally, but used less often because accurate measurement is more difficult, is the angle the celestial body makes with the vertical plane which passes through the observer's geographical meridian. This angle is called the azimuth.)

Basically, then, what a celestial navigator does is measure the altitude or zenith angle of two or more celestial bodies; record the times of his

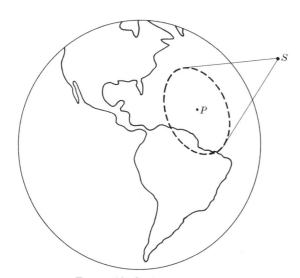

FIGURE 12. Circle of position.

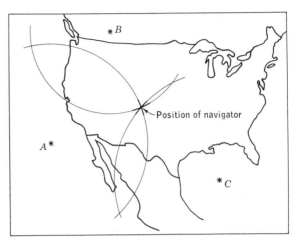

FIGURE 13. Circles of position plotted.

observations; based on these times, consult the tables of positions to
determine the points on the Earth's surface which were directly beneath
the celestial bodies at the times they were observed; mark these points
on his map; and using them as centers, draw circles of position repre-
senting the observed angles, as shown in Figure 13. The intersection
of these circles of position is the point from which the observations
were made. In most cases, observations of two celestial bodies are sufficient, since if the bodies are properly chosen, one of the two points at
which the two circles intersect is so unreasonably far away that it can
be discarded. While the procedure just described is somewhat different
than that used in the typical practical problem, it does illustrate accurately the basic principles of celestial navigation.

Satellite Navigation Systems

In celestial navigation, as we have just noted, the angle between the
vertical and the line of sight to a celestial body is measured with a
sextant or similar instrument. Position of a vehicle on the Earth's surface is determined from observations of two or more celestial bodies.
A similar procedure can be used to measure position by means of two
or more observations of a single satellite. There are, of course, some
differences between navigation by stars and by satellite, since the location of a star as viewed from the Earth changes very slowly with time

while a satellite fairly close to the Earth moves at great speed. In both cases, the observer must measure the direction of the local vertical by bubble level or some other device which provides a reference from which to measure angles.

In order to make the satellite navigation system suitable for all-weather navigation, we can have the satellite radiate a continuous radio signal. Any observer who is equipped with an electronic sextant and a vertical indicating system will then be able to determine his position from radio observations of the satellite.

It should be emphasized that many systems can be devised to exploit with precision the signals from navigational satellites. The merit of any particular system can be judged by the degree to which it makes use of the available information, the technical feasibility of its execution, and the compatibility of the technical features and functions with the operation which the system is to serve.

There are certain needs which are common to all systems. For example, all systems need an accurate clock, at least at the receiver or navigator's station. This clock must maintain universal time with an accuracy of about one-thousandth of a second since the motion of the satellite is such that it moves an appreciable distance during this time.

The navigator relying on artificial earth satellites to give him position information must be provided with a table of satellite positions covering the duration of his journey. These tables can be prepared in advance as mathematical predictions, as has been done by the Naval Observatory for many years in the case of natural celestial bodies.

Navigation accuracy, of course, is dependent upon the precision with which satellite position can be predicted. This precision, in turn, is a strong function of the accuracy of the tracking stations, the computational procedure used, the accuracy with which the relevant physical constants are known, and the magnitudes of unpredictable disturbing effects acting on the satellite. The most important of these disturbances for low-altitude orbital vehicles is the uncertain knowledge of air drag. The interactions of the vehicle and the atmosphere may be both electrical and mechanical in nature. The electrical effects, which may increase the drag over that expected from purely aerodynamic considerations, are sensitive to the charge accumulated on the satellite. Electrical drag has greater relative influence on the satellite at altitudes where gas-dynamic forces are small. The general consensus is that electrical drag contributes less than 10 per cent of the total drag of long-lifetime satellites presently in orbit around the Earth.

The orbit of Vanguard I can be predicted about a month ahead with the accuracies required for tracking purposes. Current predictions of the position of Vanguard I tend to be in error by about 10 miles after a month and by less than 5 miles over a few days. Predictions a few hours in advance are off by fractions of a mile.

In addition to errors in satellite observations, there are two sources of difficulty in trying to make accurate orbital predictions. First, the classical methods used for many years by astronomers to determine orbits have certain inadequacies when used in connection with the orbits of artificial satellites. New techniques, or modifications of existing techniques, are necessary in order to improve the accuracy of orbital predictions. Second, further study must be made of the disturbing forces influencing orbital motion.

These appear to be two particularly fruitful areas for detailed investigation by students interested in the field of astronautics. Speaking of the science of astronautics, I am a little concerned by the fact that some colleges and universities in this country seem somewhat slow at recognizing astronautics as a separate science which is on a par with aeronautics, physics, chemistry, and all the other well-recognized disciplines. I should like to encourage the recognition of this discipline by educators and students to the extent of establishing separate curricula and awarding degrees in this field. It appears to me that the astronautics curricula should extend the emphasis of aeronautical engineering in the direction of providing more mathematics, physics, celestial mechanics, general field theory, thin shell structures, feedback control theory, and gas dynamics.

The geometric problem of navigating over the Earth's surface by means of electromagnetic radiations from artificial satellites may be solved in three basic ways:

1. By angle-only measurements (altitude-azimuth)
2. By range-only measurements
3. By range-angle measurements

These correspond generally with three separate approaches to the location of a point in space (in this case on the Earth's surface) in some known reference system.

The first two of the means mentioned above have been investigated in some detail by the U.S. Navy and have been proposed as systems under the project names PATHFINDER and TRANSIT, respectively. As this is being written, TRANSIT has proceeded to the stage where we have two satellites in orbit to test the feasibility of the system. I might add that results obtained to date are very encouraging. I will now outline the fundamental concepts underlying the three basic methods of navigating from artificial satellites.

Angle-only System (PATHFINDER)

Classical celestial navigation is based on the determination of the distance OP on the surface of the Earth between observer O and subastral (subsatellite) point P which is found in tables as a function of time (see Figure 11). This determination is made by measuring the zenith angle of the celestial object and perhaps its geographic azimuth. A fix is obtained by making one or more such observations, the optimum determination being that case in which the lines of position (a line of position is a short section of a circle of position) for several observations make large-angle intersections at the location of the observer.

The accuracy of the fix depends in part on the accuracy with which the true zenith angles can be determined. For example, an error of six-hundredths of a degree in the angle of near-overhead celestial objects will produce an error of 3 to 4 nautical miles in OP. In this connection, it will be noted that the accuracy with which the zenith angle can be determined depends on the establishment of an accurate line to the center of the earth at the site of the observer. The gravitational vertical does not coincide with this line, but can be related to it by reasonably well-established corrections.

For optical observations, an artificial satellite of the Earth is, at best, less useful and less convenient for navigational purposes than prominent natural celestial objects. However, an artificial satellite can be equipped so that it constitutes an intense radio point source. With proper choice of frequency, observations can be made on such a source even when optical determinations are entirely impossible (for example, during conditions of cloud cover). Such an all-weather capability can be of enormous importance to naval and maritime operations.

PATHFINDER is an extension of familiar navigational practices. By establishing two or more lines of position from observations of the angular elevation of two or more satellites, PATHFINDER can establish a fix without delay. Even with only one satellite on which to make observations, a running fix can be derived provided the observer's motion is known.

Angles must be measured with great precision. An error of six-hundredths of a degree corresponds to an error of 3 to 4 miles in the navigational fix. In radio navigation, in contrast to optical navigation, wavelengths are relatively long and larger equipment is required to measure angles accurately. Also, there are significantly large tropospheric and ionospheric corrections which must be made. The uncertainty in measured position depends, in part, on the frequency of the signal. In order to improve angular resolution, it is desirable to use the highest frequencies that are compatible with considerations of power requirements and signal path attenuation.

The navigating station must be equipped with a stable local vertical. Local vertical must be measured to about 0.3 minutes of arc for accurate position determination; this is well within the capabilities of present gyro systems.

Range-only Navigation (TRANSIT)

The navigator may measure time intervals as is done by radar in order to determine the instantaneous range to the satellite from the point whose position is to be established. The range is equal to the time it takes the signal to reach the satellite, or to return from it, multiplied by the velocity of light. A circle of position is given by the intersection of the sphere centered at the satellite (with a radius equal to the measured range) and the surface of the Earth.

Since any direct radar system in which the navigator on the surface of the Earth illuminates the satellite and receives either a passive reflection or an immediate repetition signal from a beacon in the satellite is incompatible with military operations, the interrogation-type ranging must be rejected for a military navigation system. It is necessary to consider systems which involve only a receiver at the navigator's station. All such ranging systems, including TRANSIT, may be understood in terms of a simple model based on a satellite which transmits a sequence of accurately timed pulses.

If it were possible to provide synchronism in absolute time between the satellite and the user at the Earth's surface to about 0.1 microsecond, it would be possible to measure the true range directly by simply observing the arrival delay of pulses from the satellite. This degree of synchronism is not possible under present conditions. It is possible, however, to keep synchronism to perhaps 1 millisecond. With this degree of synchronization and with the provision that the clocks in the satellite and at the Earth's surface keep accurate rates (i.e., the discrepancy between the clocks changes so slowly that for short periods of time it can be considered a constant), the observer's position can still be determined. Referring to Figure 14, the track of the satellite can be plotted on the navigator's map on the basis of the tables of predicted satellite position. If a series of measurements (1 to 6 in Figure 14) is made of the arrival delays of the satellite pulses and from these the ranges r_{m1}, r_{m2}, etc., are computed and plotted, it turns out that they are tangent to, and form the envelope of, a circle the center of which is the observer's position. (The radius ρ of this circle is a direct function of the degree of nonsynchronism; i.e., the discrepancy between the satellite's clock and the observer's clock.) The computations must take into account the height of the satellite above the Earth's surface, this figure also being obtained from the tables of position.

The accuracy of the ranging system just described depends on the stability of the pair of clocks, one of which controls the pulses in the satellite and the other of which is used to measure r_m at the navigator's station. The nonsynchronism in time (ρ divided by the velocity of light c) must remain nearly constant throughout the series of measurements. Uncertainties in the fix are directly related to the drift in ρ/c during the course of the satellite's passage. The accuracy of the fix depends further on computational and round-off errors, and on the accuracy with which the satellite track has been predicted. In addition, there will be uncertainties arising from tropospheric and ionospheric effects which cannot be entirely corrected.

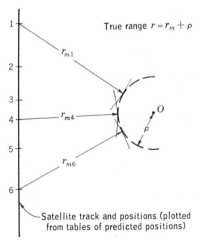

True range $r = r_m + \rho$

Satellite track and positions (plotted from tables of predicted positions)

FIGURE 14. Geometry of range-only navigation system.

In this system, the equipment required consists of a satellite-borne transmitter which delivers pulses to an antenna, the pulse timing being derived from a stable oscillator. Oscillator stability requirements are well within the state of the art of quartz crystal fabrication. The pulses delivered must be sharp, with rise times of 0.1 microsecond or better. To preserve the pulse shape and particularly to minimize tropospheric and ionospheric contributions to range uncertainties, there is great advantage in the shorter wavelengths. At 1,000 Mc, the uncertainty in the range correction for tropospheric and ionospheric effects is as large as the correction itself, whereas at 10,000 Mc, the uncertainty is trivial.

The antenna of this pulse-ranging system can be directional or omnidirectional. Directional antennas would improve the signal-to-noise ratio, but would require a satellite-borne tracking system to keep the antenna aimed at the ground station.

The navigator must have a clock with a rate derived from a stable oscillator of the same frequency as that of the oscillator carried by the satellite.

The determination of the navigator's position can be made equally well by measurement of the rate of change of range, which through proper mathematical computations can be made to yield the observer's position.

This use of the rate of change of range constitutes the basis for the TRANSIT navigation system developed by the Applied Physics Laboratory at Johns Hopkins University. This system is commonly described as a "doppler" system. The first satellite for this system was launched on April 13, 1960.

Combined Range-and-angle System

The range-only system can be supplemented with a simple cross-range angle-measuring device which will aid in resolving the cross-range ambiguities of the distance of closest approach. When this distance is small, range-only measurements become less accurate for determining cross-range position. Such a system might be justified for a large ship where the size and weight penalties of a large paraboloid reflector or an antenna array can be accepted.

Comparison of Satellite Systems

I should like to summarize briefly the equipment needed for satellite navigation systems. Basic measurements recorded in the angle-only system are elevation angles and azimuth of the satellite and the times of observation. Navigation equipment consists of a highly directional antenna and receiving system, a clock, and a vertical indicating system such as a horizon scanner or a Schuler-tuned stable platform. The accuracy of navigation is determined by various equipment characteristics, not the least of which is antenna size. Antennas can be divided into two classes, those using a single mirror to focus the energy into a small region and those using spaced antennas for interferometer measurements. The long distances traveled by the signals make it necessary for the energy-collecting area to be as large as possible. In general, the beam-width decreases and angular accuracy improves as the antenna area is increased. As an indication of this size, an antenna about 12 feet in diameter will allow position accuracy within about 1 mile when directed at a satellite 1,000 miles away under representative conditions. A smaller antenna would lead to larger errors.

Except for the antenna, other equipment required in the angle-only satellite system is relatively small and light.

For the doppler system of navigation, the physical extent of the equipment needed on the ship or aircraft is less than that required for the angle-only system. A big advantage of the doppler-shift navigation system is that it does not require determination of the local vertical by the observer. The navigation equipment consists of a sensitive radio receiver, an accurate frequency reference, and an accurate clock. The equipment carried by the satellite would again be a transmitter, but one specifically designed to operate at a very stable frequency.

We might contrast the practical features of the angle-only system versus the doppler system as follows:

1. The doppler system is potentially capable of greater accuracy.

2. The angle-only system can furnish a true azimuth from a single satellite observation.

3. Topside equipment for the doppler system is simple, lightweight, and inconspicuous. The angle-only system requires relatively large and cumbersome topside equipment.

4. The electronic equipment required in the angle-only system is simpler, since precise frequency control is not as critical.

5. The angle-only system requires a precise independent indication of the local vertical. The doppler system does *not* require a determination of the local vertical, but does require more elaborate electronic and computer installations, and there is a mandatory requirement of precise frequency control.

6. The doppler system uses satellites of relatively low orbits (700 miles or less) while the angle-only system uses higher orbits.

7. The doppler system needs a recording device by means of which it continuously transmits time signals and current orbital data, with a command-link for updating these data.

8. None of the satellite navigation systems can operate without precise satellite tracking and orbital prediction. Steps should be taken to assure the continual development of adequate tracking and prediction facilities by our government.

It might be well to review at this point some of the advantages of a navigation system employing artificial satellites over a conventional celestial navigation system.

1. The most significant advantage is the fact that the satellite navigation system can operate in fair weather or foul, day or night. It is not necessary that the navigator be able to see with the naked eye the body on which he is sighting.

2. Due to the rapid motion of the satellites, only a few are required to completely cover the areas of the Earth over which there is significant sea and air travel. There need be no problem of searching for two or more visible stars located nearly 90° apart in the heavens in order to establish a sharp point of intersection of the lines of position.

3. The satellite navigation system has considerable growth potential. By this, I mean that it has great promise of large-scale improvements, particularly in the direction of automation. Since the data from the satellite are in the form of relatively strong electromagnetic signals of radio frequencies rather than weak signals at optical frequencies, automatic tracking will be much less difficult than for optical star trackers. It is conceivable that "push-button" navigation may be obtainable without incurring a great weight penalty in terms of equipment required for data pick-up, memory storage, and computational functions.

With the progress being made today in some of the more exotic memory devices, the navigational information usually provided in the form of tables may be stored in electronic form rather than on paper. The tremendous strides being made in miniaturization of electronic switching devices, such as with superconductors and other advanced products of solid state physics, may make it possible to reduce the size and weight of the computing systems sufficiently to be competitive with other navigation concepts.

State of Navigation Satellite Systems Today

Some time ago, we in the Navy recognized the potential of a satellite navigation system, and encouraged studies and developmental programs in this direction. We have fostered research programs for both the PATHFINDER angle-only and the TRANSIT doppler-shift systems described earlier. The first two TRANSIT satellites are now in orbit about the Earth.

One of the important bonus features of these and other satellite programs is that in the process of developing the primary system, there are often important discoveries and improvements made in other fields. For example, we have made significant strides in radio sextants as part of the PATHFINDER program—this progress in turn feeds back to improve optical sextants and automatic star trackers. Improvements made in guidance systems to place the navigation satellites in precision orbits naturally feed back to improve our ballistic missile systems, man-in-space program, and guidance concepts for other astronautical projects.

As is generally the case, improvements are made across the board as a result of discoveries which grow out of research directed primarily with a specific goal in mind, such as a satellite navigation system. For example, much valuable environmental and geophysical data will result from our navigational satellite program.

The birth of the TRANSIT doppler-shift navigational system is rather interesting. It was an outgrowth of research performed by two men at Johns Hopkins University, Dr. William Guier and Dr. George Weiffenback. These men and their associates established the feasibility of the system in the laboratories and recognized the advantages of such a navigational concept. When their work was brought to our attention in the Navy, we, of course, were very interested in pursuing it further because of its great potential in sea and air navigation.

Thus, out of the academic environment, far removed from the sea and not directly concerned with or ingrained in the problems of national defense, an idea grew that will greatly benefit all traveling men. There is a vital need for close coordination and cooperation between the academic world, the research world, and the military. There is a need for a large-scale exchange of information: for a flow of requirements from the operating arena where the problems are born; and for a constant stream of ideas which may have military worth from the teacher, the scientist, and the engineer into channels where they come under scrutiny of the military.

It seems to be a rather popular pastime these days for public officials to take "pot-shots" at the nation's educational system and to play down the demands made of our students. I do not choose to play this game because I do not feel that there is anything greatly wrong with either our educational system or our students. A little more money for classrooms, laboratory facilities, books, and teachers' salaries would cure most of the ills—particularly improvement in the salary level of our teachers. I am sure that improvements can be made in the educational opportunities of our young people, but I do not feel that revolutions are required just because the Soviet Union orbits heavier satellites than the United States. It is true that some worth-while changes can be made in our educational system and in our teaching techniques, but we should be sure that we retain the good features of our old system and add only the best of the new concepts. The fact that nuclear power, inertial navigation systems, the TRANSIT system of satellite navigation, and many other tremendous scientific and engineering achievements were born and fostered in the academic environment of our colleges and universities is proof enough that revolutions are not in order.

Signpost of the Future

Space is a great steppingstone for improving man's standard of living and for broadening his knowledge of the world in which we live. I feel that the TRANSIT satellite navigational system is one of the great fundamental strides upward in the direction of utilizing space to improve the world as we know it.

It is interesting to speculate on the future of satellite systems for advancing the art of world navigation. I have talked thus far mainly about shipboard navigation and about the fundamental principles of using satellites for navigational purposes. I should like now to extend our horizon to possible applications in the air and in space.

The fundamental problems we are facing today in air traffic control are the increased volume of air traffic, the greatly increased speeds of aircraft, and the necessity for rapid decisions and actions in the event of emergencies because of these greater speeds and the greater fuel consumption. I expect to see eventually very cheap and very small air-borne inertial navigation systems—made cheap and small by the fact that they are required to maintain positional accuracy unaided only for short periods of time. There will be automatic navigational readout from navigational satellites for purposes of upgrading the inertial data. Thus, a continuous and accurate record of position information can be maintained in a minimal system.

To carry this concept one step further, it is possible for position data to be transmitted automatically at appropriate intervals, each minute or 5 minutes or so, from the aircraft navigational system to air traffic control points. Continuous and accurate control of all air-borne traffic may thus be maintained. Due to increased position accuracy and to the increased positive nature of the control in comparison to that carried out today, aircraft can be stacked closer together with increased safety.

Our space efforts today are small and well-calculated steps—much as the early steps taken by a little baby learning to walk. In time we will be on good solid ground, sure of ourselves and wise as to our objectives. Manned space exploration is bound to follow closely on the heels of our first few successful manned orbital missions. Taylor and Blockley in the book *Space Technology* have stated eloquently man's motivation for space exploration:

> The feat of putting man into orbit and the final triumph of interplanetary and even interstellar exploration are supreme human goals, transcending the purely pragmatic question of which performs more efficiently, man or machine.

A competent team of trained and motivated men cannot be surpassed by machine to perform *exploratory functions,* as contrasted to reasonably well-established *routine functions* (such as data collection and computation). The machine can do those functions for which it is designed; it cannot seek new functions beyond the scope of its sensory transducers. The design of the machine may be elaborate in order to encompass limited decision-making capabilities; the ability to adapt its performance to a changing environment; and the faculty for rapid determination, computation, cataloging, and storing of data. A machine or group of machines that can perform an almost unlimited diversity of tasks such as the following, however, will probably never be built in a package weighing less than 200 pounds:

1. Correlate many nonrelated observations
2. Readily perform multifarious physical tasks, in any order, such as moving from place to place, replacing or repairing a faulty engine or electronic component, controlling a flying vehicle, collecting minerals and plant life, comparing observations with data previously accumulated, and so forth
3. Make rapid decisions based on seemingly independent events
4. Have intuition *

Man has a unique facility for exercising judgment; he can reason inductively and has the ability to draw inferences from isolated elements of one situation and apply them to another. The trained human being has the capacity to analyze problems never before encountered and to make decisions on the basis of general rather than specific experience. He is a valuable technical trouble shooter and can better assure reliable operation of all equipment on the astronautical vehicle by continuously checking, repairing, and replacing faulty equipment. It has been demonstrated over a half century of flight that he is a versatile flight controller; there is much expert opinion to substantiate the belief that he can exercise this function to a useful extent even at escape velocities.

The planning of an astronautical mission and the design of the vehicle and its guidance and control system constitute a problem of optimum system design to effect a successful mission with maximum flexibility and reliability, minimum initial weight, and the most efficient power utilization. A realistic approach to analyzing the mission problem is to assign to humans those functions performed best by them and

---

* "Immediate apprehension or cognition; the power of knowing or the knowledge obtained without recourse to inference or reasoning; insight; familiarity, a quick or ready apprehension." (*Webster's New Collegiate Dictionary.*)

to relegate to machines the functions that they can perform more efficiently. Taylor and Blockley state man's role:

> Man will perform primarily as a strategist, as a correlator of data at many probability levels, a maker of insightful interpretations from these data, and a formulator of effective plans of action. Although it is possible to conceive of a machine that could perform some or all of these functions, it would be very large, complicated, and expensive.
>
> Man is needed as a scientific observer of space phenomena. He can act to control the instrumentation toward the most significant and reliable readings, to make the unexpected observation, and to gain basic insights into the pattern of results. Although machines could conceivably do this job, man will *want* to do the job personally, whether required or not.

Before any attempt is made to explore the surface of a strange planet, sufficient data regarding the planet must be accumulated in advance to demonstrate that the potential benefits of such an exploration are great enough to justify not only the tremendous cost but also the attendant risk to human life. Preceding the manned mission will be a certain number of unmanned exploratory probes to collect data on radiation hazards; gravity; magnetic fields; atmospheric properties such as temperature, density, and wind characteristics; terrain mapping; and other aspects of the planet. Manned exploration of the vicinity of the planet, which may or may not culminate in immediate atmospheric penetration to a landing, should follow relatively few unmanned probes.

It is interesting to compare some of the debits and credits which enter the engineering ledger as a result of designing a mission system to include a human crew. The man excels over the machine in:

1. Detecting minimal changes in visual and auditory stimuli

2. Perceiving, in the presence of noise, meaningful patterns and information

3. Choosing new courses of action with great flexibility and adaptability when circumstances change unexpectedly

4. Storing tremendous quantities of data for long periods of time and recalling the required relevant information rapidly

Engineering liabilities incurred by incorporating man in the system include the following:

1. Much engineering effort is required to make the astronautical vehicle habitable and safe due to the inherent vulnerability of man to the space environment.

2. Engineering provision must be made for living, rest, and recreation that go beyond a utilitarian minimum because more is required of the man than mere survival. He must perform at top efficiency, a requirement closely coupled with morale and physiological considerations.

3. Equipment required to make man effective in his duties, such as human-operator stations, data-processing and storage facilities, sensory displays, and the like, is costly in weight, size, complexity, and dollars.

The net cost of each unmanned planetary probe, in terms of man-hours and natural resources expended versus data collected, is sufficiently great that emphasis should be placed on the manned mission as early as possible. The potential wealth of knowledge that may be obtained versus expenditure of resources and money is greater, by orders of magnitude, in manned exploration as contrasted to unmanned probes, even though each manned vehicle must itself be more elaborate and costly. Therefore, there is strong justification for manned exploration to follow relatively few unmanned probes. The latter should be undertaken in such numbers only to establish that the risk factor to man (such as from radiation and meteor hazards, launch and recovery failure, and operational reliability) is within acceptable bounds.

First-time entry into the atmospheres of strange planets presents special problems in specifying position when compared to navigating over a well-mapped planet such as Earth. The choice of a suitable landing site must necessarily be based on reconnaissance of the planetary surface while in orbit around the planet. The orbital altitude for the reconnaissance phase must be high enough so that a prolonged orbit may persist, yet low enough that fairly accurate mapping of the terrain is feasible.

The navigation of a vehicle flying from one point to another point on the surface of a planet is generally based on navigational parameters measured with respect to the planet; i.e., latitude, longitude, and altitude. A less common method is to use parameters identified with a particular mission, such as angular displacement measured along-track and across-track with respect to a great circle course.

The entry mission to the surface of a strange planet, on the other hand, does not originate from a point on the planet's surface. The entry mission generally originates from a reconnaissance orbit which, if the perigeal altitude is sufficiently great, is only very slowly changing with respect to an inertial framework. This change is caused by small horizontal components in the planet's gravitational field. Drag forces result in energy transfer from the vehicle to the planetary atmosphere, but for sufficiently high orbits, this transfer causes negligible change in the satellite orbit over periods of time comparable to that required for entry once retro-rocket thrust is generated.

A basic position reference available during the course of entry is the original reconnaissance orbit. If the entry vehicle is launched from a mother satellite, then the mother satellite, which remains in the reconnaissance orbit, may track both the entry vehicle and the pre-selected landing site and transmit this tracking information to the guidance computer of the entry vehicle. In this way, the parent satellite replaces the ground tracking stations which are used as a source of tracking information for Earth satellites and entry vehicles.

In the event that no parent satellite exists, then a navigational satellite to serve the same purpose may be deposited in the reconnaissance orbit by the entry vehicle prior to initiating the entry phase.

In such a system, the landing site may be considered to be a target moving in three-dimensional space with respect to the near-stable reconnaissance trajectory of the mother or navigational satellite. The entry vehicle is also moving with respect to the reconnaissance trajectory. The entry problem is therefore similar to the fire control problem, with the entry vehicle (projectile) fired from the parent satellite (gun) to hit the moving landing site (target). The problem is much more severe than the conventional fire control problem, however, because the entry vehicle must be constrained to paths in which it will not burn up or encounter accelerations beyond tolerable levels.

All space programs require large boosters, lots of money and man-hours, and closely coordinated efforts. The navigation satellite program, since it is primarily a nonmilitary effort, is an area in which the United States and the USSR could very readily cooperate to the mutual benefit of both sides concerned—not to mention the benefits spread to all the rest of the world. International cooperation in astronautics is necessary as a matter of efficiency. Scientific space exploration cannot easily be carried out in isolated national packages. The history of astronomy as an

84 VICE ADMIRAL JOHN T. HAYWARD

international science demonstrates the fact that some scientific disciplines do not readily recognize state boundaries. Observation of natural celestial bodies has required the closest kind of international collaboration. The creation, observation, and retrieval of artificial celestial bodies places even more urgent demands on international cooperation. There is an obvious need for international cooperation in such matters as agreement on radio-frequency allocations for space vehicles, and on rights of access to (and exit from) national territories for recovery of vehicles. Astronautics raises substantial questions of law, both international and local. The important issue of international agreement on space exploitation must be afforded serious consideration.

Although astronautics is inherently a high-cost activity that will clearly have an important impact on government expenditures, taxes, corporate profits, personal incomes, and our military posture, I feel that its future holds considerable promise of substantial economic benefits. The satellite navigation system is an early instance of a space system that gives promise of more than paying for itself in terms of net return in money, conservation of man-hours, and personnel safety.

5

Application of Space Science to Earth Travel

LESTON FANEUF

FORMERLY CHAIRMAN OF THE BOARD AND PRESIDENT

BELL AIRCRAFT CORPORATION

Leston Faneuf is a man of varied backgrounds ranging from teaching, journalism, editing, and banking to management of a major aircraft company.

On being graduated from Norwich University in 1926 he became commandant of DeVeaux School, Niagara Falls, N.Y., and later an instructor in Nichols School, Buffalo, N.Y.

From 1933 until 1943 he was a political editor, radio news commentator, assistant vice president of the Marine Midland Corporation, and secretary to the mayor of Buffalo. Several of these positions were held concurrently.

Faneuf joined Bell in 1943 as assistant to the president, becoming general manager in 1954; president in 1956; chairman of the board and president in 1958, and chairman of the board in 1959. In 1960 he retired from active Bell management and is presently serving as a director of Bell Aerospace Corporation.

Faneuf is a member of the National Board of Directors of the Crusade for Freedom, and the American Red Cross; a director of the Associated Industries of New York State; vice president of the National Defense Transportation Association, a director of Marine Midland Corporation and of the Western New York Nuclear Research Center.

His honorary degrees include: Doctor of Laws, Colgate University, 1957; Doctor of Engineering, Clarkson College of Technology, 1957; Doctor of Science, Canisius College, 1958; Doctor of Commercial Science, Ithaca College, 1958; Doctor of Laws, Alfred University, 1958; Doctor of Science, Norwich University, 1959.

ONLY A few short years ago—in 1943—the first American-built jet airplane, the Bell P-59, was flown at Muroc Lake, now Edwards Air Force Base, in California. Yet today, the jet age is full upon us; public acceptance of our first commercial jet passenger transports has been overwhelmingly enthusiastic. As we casually shuttle across a continent or an ocean at 600 miles per hour and at an altitude of 7 miles, few of us bother to consider that we are flying faster and higher than most fighting planes of World War II.

Such is our complacent acceptance, and almost casual utilization, of the tremendous technological advancements in aeronautics during the two decades from 1940 to 1960. But, during the latter of these two decades, the science of aeronautics evolved into astronautics. The jet age had scarcely arrived before it was evolving into a new and incredible era—the space age.

The space age actually dawned on the general public with something like explosive impact on October 4, 1957—the day of the Sputnik. Of course, scientists and professional airmen had been more and more aware of its approaching possibilities as a result of their postwar rocket and missile development work. Also adding to technical awareness of the approaching space age were the significant flight histories of two U.S. Air Force research airplanes, the Bell X-1 and X-2. These planes first pioneered through the sound barrier and then to a height of 126,000 feet at more than 2,100 miles an hour, the world's highest and fastest manned flights until 1960.

Led largely by transplanted German V-2 rocketeers like Dornberger, von Braun, and Ehricke, government and industry research departments were gazing at heights far beyond the X-2's 126,000 feet and even beyond the then drawing-board X-15's 100 miles of altitude capability. The Dornberger hypersonic boost-glide concept that eventually spawned the Dyna-Soar project had been under study since the early 1950s, but only a few hardheaded believers like Gen. Curtis LeMay insisted on continuation of man-in-flight, despite the glamorous promise of the automatic missile age. Hence, enthusiasm and support in high places for such manned space vehicles as Dornberger's were thin and tepid.

The Soviet Sputnik changed all that! Space suddenly opened its vast reaches with inviting arms, as if the 100 billion stars around our small earth had just been switched on like a Christmas tree display.

Our domestic controversy during the presidential election year of 1960 obviously fogged many of the facts in the debate on the U.S.-Soviet rocket and missile gap. But, I think most thoughtful arms-length observers agree that, regardless of any relationship to Soviet technology, our own scientific progress in this country in the fields of rockets and space research has been tremendous in the past 10 years.

And this brings us to the subject of this chapter. Let me state it this way: Less than 18 years after the first U.S. jet airplane made its initial flight, thousands of passengers daily are riding jet transports all over the world. Will the same public utilization be made 18 years from now of rocket and space techniques currently under development and sure to be developed further in the years immediately ahead? Public utilization, not to fly to the moon and back, not for fantastic interplanetary travel, but simply to travel from one spot on the earth to another spot in an elapsed time of perhaps 1 hour?

Before we can hazard a categoric "yes" or "no" answer to this question, we might examine some of the factors involved:

1. What do we expect will be the evolution of so-called conventional air transports in the next 20 years?

2. Will space techniques and propulsion systems be feasible and safe for commercial transportation by the 1980 to 1990 period?

3. What about the economics of such transportation?

4. Will the public, or at least some substantial segment of it, be ready and willing to utilize a hypersonic transport?

Incidentally, we will be discussing supersonic transports of 1975 and hypersonic transports of the 1980–1990 era. The supersonic airliners are assumed to be flying at a velocity of 1,500 miles per hour and the hypersonic airplanes at about 15,000 miles per hour.

As we look for the answers to our four basic questions in an effort to find a single conclusive answer to our hypersonic flight query, we should emphasize that this is not primarily a technical paper. It is designed, rather, to take a practical business look at possible future commercial travel applications of guided missile developments and space technology, obviously giving close scrutiny to the present-day state of the art.

Now, what do the transport people have on their drawing boards for the next generation of commercial transports, following today's jets? Most United States companies in the transport manufacturing field are obviously thinking in terms of supersonic jet airliners that will fly 1,500 to 2,000 miles per hour on the world's longer routes.

And let us not forget that our British friends first showed us the way in jet engines, turboprop transports, and jet transports. The British government has asked two new airframe groups and its two engine groups to make joint design studies for a supersonic airliner, according to an announcement made in 1960 by Duncan Sandys, Minister of Aviation.

Therefore, it seems reasonable to assume that we can envision in the early 1970s a variety of supersonic jets being readied to replace, on longer routes, our present-day 600-mph jet transports. The economics of operating airlines will be the factor determining the precise dates that these supersonic airplanes will start replacing or supplementing the British Comets, Boeing 707s, Douglas DC-8s, and Convair 880s.

Now, if these supersonic jet transports will be capable of carrying you 6,000 miles in 6 hours, including a refueling stop, do you want to fly faster, and if so, why?

Suppose we look at how much faster we think we can travel before we answer. The space sciences, learned in the pursuit of military objectives and demonstrated in laboratories and in hypersonic wind tunnels, permit us to foresee vehicles to carry passengers in the fringe of outer space. Dr. Walter Dornberger of Bell Aircraft Corporation predicted in 1958 that rocket-powered, commercial airliners, flying above an altitude of 150,000 feet and at speeds exceeding 11,000 miles per hour, would one day be practical products of the space age.

No one today can be brash enough to furnish pat answers on the problems of structure, propulsion, heat, and aerodynamics that will play such a vital part in our space thinking and space "doing" 20 years from now. Everyone in the space field is keenly aware of what *Aviation Week* magazine has pointed out; namely, that it is almost impossible to find agreement among hypersonic research scientists as to the ideal configuration for a hypersonic aircraft. Also, precise knowledge of space environment is uncertain at present; however, we believe that we have means to protect man from this environment, hostile as it is. But to bring a controlled vehicle back from space is yet to be demonstrated under conditions suitable for manned flight. Continued research and development work in heat-sink structure, double-wall construction, and super-alloys will solve the re-entry heating and the deceleration of a space glider. The propulsion systems used in supersonic military and research aircraft and those demonstrated in launching space vehicles will·provide an adequate background for choosing the propulsion units for a space age transport.

Let's review the technical assumptions leading to a boost-glide rocket transport. The initial velocity may be attained by taking advantage of a one- or two-stage booster, the same general principle being followed in our ICBM missiles. The upper limit of our velocity must necessarily be held below 18,000 miles per hour, since at this speed the vehicle would be established in a rather permanent orbit, which could disrupt our travel time, to put it mildly.

These and hundreds of other accomplishments from the space flight work of NASA, the military, the research institutions, and the universities will furnish sufficient data for our space-minded industry to design a hypersonic transport. Such a transport would permit travel between continents at an average speed of 7,000 miles per hour.

The so-called boost-glide principle enables us to obtain a long-range flight with a minimum time of applied energy. This means that, in contrast to the supersonic airplane with its constant high thrust, our booster thrust is applied only during the first few minutes of flight. After all the thrust has been applied and maximum speed is reached, the free fall and glide of the vehicle dissipates the initial energy over a 1-hour period and results in near-global ranges.

The first-stage booster could be either of two configurations:

1. A vertical take-off rocket booster which necessarily has high accelerations, uncomfortable to the passengers, and requires a complex launching apparatus to hold the booster upright on its tail, ready for the initial vertical trajectory

2. A horizontal take-off, similar to that employed today by conventional jet transports, which will benefit by obtaining considerable lift from airfoil surfaces and will have relatively low acceleration

The booster used for such conventional take-off will have the additional advantage of a landing gear for use on both take-off and landing. Such a conventional take-off means that the booster and vehicle itself can conceivably be operated from conventional 1980 airports, thus saving the elaborate launch complex usually required by vertically ground-launched ICBM's. For these reasons, the horizontal take-off appears to be the better of the two configurations.

For economic reasons, we should also eliminate the extravagance of expending or dropping propulsion systems after take-off. So, it will be desirable to recover the first-stage booster by having a pilot fly it back to the take-off site. Also, we will retain the second-stage power plant as an integral part of the transport airplane itself. While we are planning an air-breathing or jet booster for the first or take-off stage, the airplane power plant must be a rocket because of the high-altitude operation which will be above air-breathing capability. The major operating cost will be for propellants for the air-breather jet and for the rocket, and not for hardware used once and then discarded as a total loss.

Figure 15 shows our hypersonic transport mounted piggy-back fashion on the air-breathing booster. The transport fuselage is 88 feet long, and the front portion of it is a cockpit and a cabin containing reclining seats for some 30 passengers.

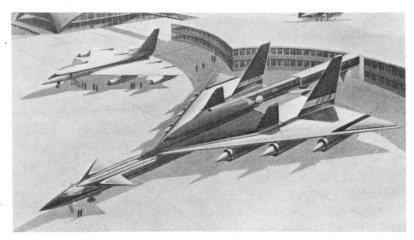

FIGURE 15. Hypersonic transport mounted on air-breathing booster.

The forward cockpit of the transport houses a crew of two and is equipped for blind flying. A sliding heat shield can be opened to permit the pilots a clear view forward during approach and landing, but this shield remains closed during the rocket and glide phases of flight. The high temperatures encountered during re-entry into the lower atmosphere would make any windshield incandescent and useless as a viewing window.

A gimbaled rocket engine provides the final stage thrust. Storable liquid propellants burned for a brief interval after separation from the booster accelerate the transport to near-orbital velocity of 15,000 miles per hour. The propellant tanks occupy the rear half of the fuselage of the transport. The propellants can be pentaborane (B_5H_9), hydrazine (N_2H_4), nitrogen tetroxide (N_2O_4), or more-efficient and more-economical fuel-oxidizer combinations that may be available by 1980.

A sufficient quantity of these propellants is carried to provide 165,000 pounds of thrust for approximately 4 minutes. This thrust approximately equals the weight of the transport at separation from the booster and reaches a maximum of three times the weight at final burnout. Thus, the passengers experience a maximum of three g's pushing them into their comfortable, form-fitting, reclining seats—certainly not an excessive force.

As the transport re-enters the lower atmosphere during the latter stages of glide, its highly swept delta wing provides aerodynamic lift. The leading edges of the wing and the nose of the fuselage are covered with refractory material that can withstand temperatures up to 2500°F. The remainder of the wing and the fuselage surfaces are covered with a double-wall type of heat shield. This shield consists of rectangular panels made of super-alloy sandwich construction. These panels are insulated and stand off from the main load-carrying structure which may be liquid-cooled. Because of this heat protection, the internal structure may be made of conventional materials. In fact, with the addition of water-cooling, this structure can be aluminum alloy.

The remaining systems for the transport are designed to withstand the unusual environment of space travel. An air-conditioning system with refrigeration equipment maintains the cabin at normal ambient temperatures while the outer skin glows red hot. An inertial guidance system provides automatic control during all stages of flight.

The sequence of events and the demands for attitude control will approach the limit of pilot capability. An on-board computer can determine the flight path and provide the energy-management to obtain the desired range. A reaction control system will be required to obtain proper re-entry attitude because aerodynamic forces are insufficient during flight above the atmosphere. This system could use small liquid-propellant rocket chambers located in the nose and wing tips.

The weight of the transport at landing may be as low as 55,000 pounds. However, the total gross weight at take-off will approach 160,000 pounds because of the rocket propellants. The recoverable air-breathing booster, which will carry the transport aloft, has a fuselage approximately 200 feet long. This slender shape is only large enough in diameter to contain the fuel for the short booster flight. The two-man crew, located in a forward cockpit, monitors the automatic flight control system during boost and then pilots the booster back to the take-off airport in a conventional manner.

The delta wing of 150-foot span and high-sweepback angle supports six turbo-ramjet engines. These engines develop 50,000 pounds thrust each at sea level and accelerate the booster to 5,300 miles per hour at 120,000 feet. The turbo-ramjet to be used in the air-breathing booster will consist of a supersonic turbojet in front, an air bypass around the turbojet, and a ramjet located behind. The turbojet is used for initial thrust at take-off and climb. At some supersonic speed the inlet to the turbojet is shut off and the air bypasses the turbojet and permits the ramjet to carry on. Such engines are being proposed by two companies in this country.

Our hypersonic transport is mounted to the top of the booster fuselage on two rails. The entire exterior of the booster may be covered with titanium skin insulated from the main structure. This protects the booster body both from the heat of supersonic flight and from the blast of the transport rocket as it leaves the rails during air-launch. High-speed aerodynamic heating on the air-breathing booster can be accepted largely because of the short duration of the high-speed burst.

It seems reasonable to assume, then, that in the 1980 to 1990 era, space techniques can be made applicable to boost-glide transports by the time the airlines are ready to use them. We foresee no need for a major technical break-through, but only continuing space developments in the structures, propulsion, and aerodynamics fields. This may seem at first glance like an oversimplification. But, while recognizing the technical magnitude of developing and building a hypersonic transport, we do believe there is ample progress being made in missile and space work to justify our assumption: that man will be able to travel commercially at hypersonic speeds some 20 to 30 years from now. Obviously, the economics of developing, building, and operating such vehicles must be proved sound and practical.

Let us look at these economics and what they indicate. First is the old familiar question: What will it cost? The major costs for a hypersonic transport break down into:

1. Research and development
2. Production cost of airframes for booster and transport
3. Operation, including fuel, maintenance, and crew

The research and development required to bring the hypersonic transport into being will require at least 15 years of concerted effort. The aerospace industry is ready, willing, and able to accomplish the work. However, it is probably unable to shoulder the full financial burden. Certainly, industry cannot afford the extravagance of duplicating and competing in hypersonic designs. A coordinated team effort similar to the Air Force B-70 development program might well pay off in cost and time.

Fortunately, space-flight projects such as Dyna-Soar will cover the problem areas of stability, control, and re-entry, and the solutions will be directly applicable to the hypersonic transport. Fortunately, too, the supersonic bomber projects, such as the B-70, will accomplish much of the research and development for an air-breathing booster. The National Aeronautics and Space Administration's Project Mercury likewise will provide much of the research and development required in environmental control, global tracking, and communication for the hypersonic transport.

The total research and development bill may exceed 1 billion dollars; however, governmental support of the programs mentioned and the 15 years over which it is spread may make the cost tolerable.

The initial hardware cost of the boosters and the hypersonic transports is another big item. Can operational revenues cover the deprecia-

tion of the equipment and the operating cost and still produce a profit?
If it can, then we should probably consider that the hypersonic transport
be owned and operated by commercial airlines.

Since costs are related to weight, a weight summary is given in
Table 1.

TABLE 1. WEIGHT SUMMARY

Weight, lb

Hypersonic transport
Airframe (including rocket engine and tanks)   47,000
Propellants ($N_2H_4 + N_2O_4$ for 6,000-mile trip)   105,000
Payload (passengers)   8,000

    Subtotal   160,000

Air-breathing booster
Airframe   236,000
Fuel (for boost and return)   354,000
Payload (hypersonic transport)   160,000

    Total   750,000

The total take-off weight of the booster and the piggy-back transport
is 750,000 pounds, only 50 per cent more than today's heavy bombers.
To estimate the cost of the air-breathing booster, we note that its
weight alone is estimated at 236,000 pounds. A reasonable cost of supersonic aircraft for booster service is assumed to be $67 per pound. This
gives us a booster unit cost of $15,800,000. If this initial cost is pro-
rated over 10,000 flights, say, the cost is approximately $1,580 per flight.
The propellants required for one air-launch and return to the airport
are estimated at 354,000 pounds. At a cost of $0.02 per pound for jet
fuel, we must allow $7,080 per flight. Other costs for the operation of
the booster such as maintenance, insurance, taxes, and crew salaries
are estimated to be $1,000 per flight.

The weight of the hypersonic transport airframe as distinguished
from the booster airframe is estimated at 47,000 pounds. Here, a reasonable cost is assumed to be $100 per pound or an initial cost of $4,700,000
per ship. This cost, if amortized over 5,000 flights, amounts to $940
per flight.

Propellant cost, the major part of the operating expense, is prohibi-
tively high at present. A possible second-stage propellant combination
and the present costs are shown in Table 2. It is obviously difficult to

estimate the costs of propellants 25 years in advance, but we can note that the cost of some propellants (i.e., liquid oxygen and kerosene) has been reduced to $0.02 per pound by mass production. We will compromise and use an average cost of $0.06 per pound.

TABLE 2. PROPELLANT COSTS

Fuel	Hydrazine	N_2H_4	$2.00 per lb
Oxidizer	Nitrogen tetroxide	N_2O_4	$0.06 per lb

Since the rocket propellants weigh 105,000 pounds, this means a cost of $6,300 per flight for second-stage rocket propellants. Other cost assumptions for the operation of the hypersonic transport, such as maintenance, insurance, taxes, and crew salaries, total $2,000 per flight.

All the estimated costs are summarized in Table 3 on a per-flight basis. Airframe amortization, fuel, and other costs are $18,900 per flight.

TABLE 3. COST SUMMARY

	Cost per flight
Air-breathing booster	
Airframe	$ 1,580
Fuel	7,080
Other costs (crew, etc.)	1,000
Hypersonic transport	
Airframe	940
Propellants (6,000-mile trip)	6,300
Other costs	2,000
Subtotal	$18,900
Profit	3,000
Total	$21,900

Adding a reasonable profit figure of $3,000 per flight, we will need a total revenue of $21,900 on each trip. Based upon 30 passengers, this is a fare of $730 per passenger for 6,000 miles, certainly within the realm of economic feasibility.

It is interesting to note that a range extension of 2,000 miles requires added rocket burning time of only 10 seconds. This costs approximately $20 per passenger and thus, as shown in Table 4, the fare changes only slightly from the shortest to the longest range.

TABLE 4. FARE SUMMARY

Range, miles	Fare	Fare per 1,000 miles
4,000	$710	$178
6,000	730	122
8,000	750	94
10,000	770	77

So much for the new equipment costs and operational costs. An operator or group of operators may need 20 boosters and 40 transports to cover a 10-city intercontinental route network. The total cost of $504,000,000 for the fleet is tabulated in Table 5. Even in the expanding economy of the future, this will require an energetic financial effort but probably is not an impossible goal.

TABLE 5. TOTAL FLEET COST

Transports	$ 4,700,000 × 40 =	$188,000,000
Boosters	15,800,000 × 20 =	316,000,000
Total		$504,000,000

One solution is to consolidate into joint ventures for our hypersonic intercontinental service. This could eliminate route and schedule duplication. Another benefit of joint ventures is initial equipment financing, with more capital available for investment in a single program. Any business group interested in financing our new mode of transportation is immediately confronted with three problems:

1. What amount of money is needed to finance the hypersonic transport? The development cost, initial equipment cost, and operating costs have been estimated.

2. Will money invested return a profit? Here the costs must be balanced against the revenues.

3. What is the most appropriate means of raising the money? Both conventional and possible future sources must be investigated. Public stock offerings may not suffice in this case. The magnitude of development cost, the risk, the competition, and the international aspects will be different from the present situation.

The recovery of development cost is difficult to assign. In 5 years we may have a new propulsion system which would make obsolete the one proposed here. On the other hand, our hypersonic transport and air-breathing booster types may be usable for 20 or 30 years if no replacement comes. With this uncertainty, the risk is high and the period for recovery is unknown.

How do we plan to recover such a large development cost? Perhaps we don't. The government may have to finance such a program by relating it to military development and to international prestige. Such subsidy of development in transportation is not new to us.

Let us note the past pattern of introduction of new air transports. The United States excelled in the reciprocating-engine airliners (i.e., DC-3 through DC-7, and the Convair and Constellation). The British took the lead in turboprops with the Viscount and in turbojets with the Comet. The United States, however, regained the lead with production of the 707, the DC-8, and the 880; no doubt the British may try to regain some advantage by being first with a supersonic jet. If this sequence continues, the United States could be expected to come back with the hypersonic transport. That is, unless we cannot afford it. Ironically, the Soviets may enter the race as a supplier of the modern transports of the future. And they can afford it, if they decide to do so.

Expansion by combination of airline operations would appear to be desirable for our future hypersonic financing and for reducing operating competition in a limited market where profit is sensitive to load factor. There is a chance of public opposition to moving toward monopoly or concentration of economic power. The airlines, however, would not be reduced in number, but would be combining for financial strength to provide a better service. Surely, such an economic innovation is no more drastic to contemplate than some of our present-day political combinations in international relations.

Only recently, W. A. Patterson, president of United Airlines, warned that by 1961 mergers within the airline industry will be the only solution to the financial difficulties many carriers will face as a result of growing competition for traffic in major markets. At the same time, Patterson predicted that supersonic jet transports were at least 12 years off because, in his opinion, technological problems and heavy financial undertaking stand in the way of more immediate development of such aircraft. The financial problem would appear to be the dominant one, however. And after the supersonic jet does come, the same considerations quoted by Patterson will play an important part in determining the exact dates that hypersonic transports, such as the one we have described, will be developed and put into operation.

As far back as 1956, Dr. Robert Cornog of the then Ramo-Wooldridge Corporation had a two-installment paper in *Aeronautical Engineering Review* on the economics of rocket-propelled airplanes. And, although at that time we certainly were not technologically as far along the space path as we are now, Dr. Cornog made a strong case for the economic feasibility of hypersonic rocket-propelled passenger transportation. He wrote:

> In most parts of the world it is cheaper to buy, maintain, and service an oxcart than it is an airplane. In spite of this fact, there are more airplanes than oxcarts in most civilized areas of the world. The key to this paradox is the relative speed of transportation afforded by each vehicle.

While this Cornog observation may seem a bit facetious, he does conclude in all seriousness that under conditions and assumptions outlined in his two articles, "the rocket-propelled airplane can be made to yield profit at a rate that is substantially greater than that obtainable with any competing subsonic airplane."

Assuming, then, that a rocket-propelled airplane can be developed and built to be an economical form of earth-bound transportation, utilizing routes through the lower reaches of space, we are brought to an abrupt halt when we ask the next question: "Why does man want to travel at an average speed of 7,000 miles per hour? Won't the 1,400 miles per hour average of the supersonic jet of the future be fast enough?"

To provide a rational answer to these questions, we must look at such diverse factors as man's motivations; the probable state of international affairs both commercial and political in the years 1980 to 1990; and the economic incentives for investors, developers, and transport operators.

Former President Eisenhower has repeatedly spoken of "the growing interdependence of nations." This is probably one of the significant factors that will demand, from time to time, the speediest intercontinental means of passenger travel that science and technology can provide. Without doubt, administration of future intercontinental and international functions will require frequent and rapid exchanges of official visits. International administrators may have to address legislatures of half a dozen countries in rapid succession to obtain necessary political actions. United Nations committee hearings by that time could well be world-wide extensions of our present-day congressional hearings. Rapid intercontinental transit of committee personnel and committee staffs will be highly desirable—so will frequent and immediate consultations among high government officials in the international control of communications and transportation.

Despite our continuing hopes for peaceful settlement of international squabbles and incidents of unrest, speedy intercontinental visits by United Nations officials will be required. Perhaps, before the end of the century, international candidates for international offices will want to conduct international campaign tours on high-speed schedules to speak to international voters. And if we may philosophize a bit at this point, it might be that high-speed intercontinental travel based upon this concept of interdependence of nations could make a substantial contribution to the cause of enduring world peace. If this should materialize, even to some small extent, it could well be one of the soundest investments ever made. If this political interdependence continues to grow, as seems certain, business and technical interdependence will inevitably follow. News coverage, the medical sciences, and even pleasure travel will want the new, short travel time between continents.

Having made this much of a case for our rocket transport, I would like now to examine some of the problem areas that exist. These items will challenge our engineering talent for a few years, but I am confident each will be conquered in time for its application to our future travel needs.

Traffic control, already overloaded, will require a major revision to take care of the numerous feeder helicopters, vertical take-off jets, robot freighters, supersonic airplanes, and space transports. Fortunately, some government agencies are already thinking about this. The Air Transport Association has aptly described the solution in this manner:

> The facilities expansion and the research and development program eventually will culminate in a completely automated system of air traffic control. To get a take-off clearance, a pilot will simply push a button and the controller will respond with a signal which will provide visual clearance information on a cockpit display. En-

route, the pilot will make his position reports by push-button. The information, fed through the data-processing equipment, will be on visual display in a system of colored lights before the controller. When changes in flight path become necessary, the controller will transmit the information to the cockpit display by push-button. Automatic approach control will guide the pilot, through signals to his display panel, to an airspace "ramp" for landing.

In 1975, the airspace near the surface probably will be a maze of helicopters and small private airplanes carrying us for work and pleasure. Above this, at altitudes from 1 to 12 miles, we will have 10 layers of jet-propelled planes crisscrossing the continents. Overhead there will be a hundred satellites, some dipping as low as 100 miles. However, the space between, from 12 to 100 miles, will be relatively free of traffic. The ideal altitude for the boost-glide transport is within this range. Ninety-five per cent of the trajectory is at 12 to 60 miles altitude. Confining a large part of the intercontinental travel to these altitudes will reduce the congestion in the air and provide a corridor safe from collisions, weather, and adverse air currents, and still be below the radiation belts. Exploitation of this unused airspace seems justified if only from safety and traffic control standpoints; however, it is fortunate that the performance requirements of the boost-glide transport also fit into this corridor.

A second and obvious problem is controlling the noise level during take-off. Sound level produced by the air-breathing booster is proportional to thrust and inversely proportional to distance. A full evaluation of disturbance, however, involves many other factors such as runway utilization and relations with communities involved. The standard set by the Port of New York Authority as the upper tolerance limit is 112 perceived noise decibels. This occurs at 1,700 feet distance from a Boeing 707–120 jetliner at 12,000 pounds thrust. A 50,000-pound-thrust engine would create approximately 138 perceived noise decibels at 1,700 feet. This is 18 decibels over the upper tolerance and must be attenuated by distance. This suggests a 5-mile-long uninhabited area for the airport. Incidentally, this is just what is proposed at New Jersey's Morris County airport location by the Port of New York Authority.

One-hour intercontinental flights would be useless if we had to spend several times as long getting to the airport, which brings up the third problem, rapid feeder transportation. Cities such as Los Angeles and San Francisco, now contemplating spending hundreds of millions of dollars on rapid transit, obviously will consider connecting their urban transportation systems with their airports. Feeder service to airports out several hundred miles can very well be done with larger helicopters and vertical take-off and landing jets. This service should be directly from centers of business or residence to the intercontinental airport.

In summary, let us note that the state of the art in the space sciences is far enough advanced to predict that the present-day transports will first evolve into supersonic transports, and then into hypersonic transports, both in the foreseeable future. A recoverable air-breathing booster, together with a second-stage rocket, can impart sufficient energy to a vehicle to enable it to glide halfway around the world. We are confident that technically feasible solutions can be devised for each of the problem areas: noise, propulsion, aerodynamic heating, and traffic control. The economic justification seems to be dependent upon governmental support of the 1 billion-dollar development cost. Yet, the cost and revenue estimates show that airline operation as joint ventures among airlines can be profitable. The sociological desirability of using space paths for earth travel will require general public acceptance and favorable international relations; however, the basic motivations of man will naturally promote the conditions for acceptance.

We can say that transportation at hypersonic speed is amenable to the application of space technology. The challenge is to the space scientist, the engineer, the aerospace industry, the airline operator, the financier, the government, and the public, who, in the last analysis, are the government.

We should support the airline jet replacement program to maintain a healthy economic climate in air transport, extending government support of the supersonic bomber program as development background for an air-breathing booster. We must continue space sciences work leading to demonstration of the gliding re-entry concept, and we must promote public acceptance for these new modes of rapid transportation.

Following are figures which illustrate a typical boost-glide trajectory profile, comparison data, transit times, and illustrations of the various phases of a hypothetical trip from Los Angeles to Paris.

Figure 16 shows the altitude-range profile for a typical 6,000-mile boost-glide flight. A conventional take-off of the booster will require

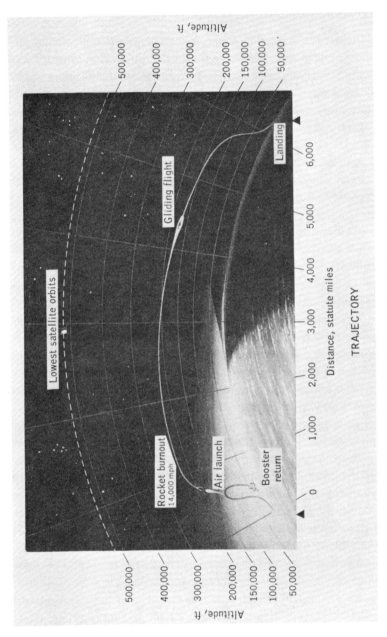

Altitude, ft

500,000
400,000
300,000
200,000
150,000
100,000
50,000

Gliding flight

Lowest satellite orbits

Rocket burnout
14,000 mph

Air launch

Booster
return

Landing

6,000

5,000

4,000

3,000

2,000

1,000

0

Distance, statute miles

TRAJECTORY

Altitude, ft

500,000
400,000
300,000
200,000
150,000
100,000
50,000

FIGURE 16. Typical altitude-range profile for hypersonic transport.

approximately 5,800 feet of runway; thus, the use of conventional airports will be practical. Climb-out can be at a 15° angle. This will enable the pilot to obtain 5,000 feet altitude as he leaves the 5-mile noise isolation area. A subsonic climb will be made to 40,000 feet. However, the take-off will be confined to a restricted corridor to avoid other aircraft. If possible, it will be over water or other isolated areas to reduce the noise problem. The remainder of the climb to 120,000 feet will be supersonic and directed toward the destination. Booster cutoff will be at a velocity of approximately 5,300 miles per hour. At this point, about 10 minutes after take-off, the rocket engine will be started and the transport separated from the first-stage booster. The transport will travel a short distance along the rails on top of the fuselage of the booster and will climb away at a shallow angle. The booster will gradually decelerate and the crew will return it to the airport, using cruise power on two of its engines.

The transport engine, burning its storable propellants, will accelerate from 5,300 miles per hour at separation to 15,000 miles per hour in order to obtain sufficient energy for a 6,000-mile glide. At this speed, centrifugal force provides approximately 72 per cent of the lift. Burnout altitude will be approximately 43 miles high. The inertial guidance system will furnish range and altitude control so that the energy is dissipated to obtain the desired range.

During re-entry into denser atmosphere, the automatic pilot will program the angle of attack to minimize the aerodynamic heating on the nose and on the leading edges of the wing. When re-entry has been completed and the airplane has returned to the lower atmosphere, the pilot can check his navigation and make corrections. When he slows to subsonic speed, a small jet engine in the transport will be started to permit him to cruise overland a short distance to the airport. The landing approach and touchdown at 200 mph will be automatically controlled and the pilot need only monitor this operation. The total elapsed time from launch has been approximately 1 hour.

Table 6 shows a comparison of speed, range, and cost of air transportation. Several assumptions have been made to simplify these numbers and a general agreement has been reached with the estimates of other investigators.

The average speed of present-day subsonic jets is approximately 500 mph. The supersonic jet of the 1970s will probably have a velocity

TABLE 6. COMPARISON DATA

	Present subsonic jet	1975 supersonic jet	1975–2000 hypersonic glider
Average speed, mph.........	500	1,400	7,000
Economic range, miles.......	4,000	5,000	4,000–11,000
Average fare per 1,000 miles..	$115	$150	$178 to $77

significantly faster and, for comparative purposes, it has been assumed to average 1,400 mph. The hypersonic glider which reaches a maximum velocity of over 15,000 mph will average approximately 7,000 mph over long-range flights.

The range of present-day jets is about 4,000 miles. The supersonic jet will be practical to 5,000 miles. Our hypersonic glider, on the other hand, will be impractical for ranges under 4,000 miles. This immediately eliminates it from the cross-continent transportation system. The ultimate range of the hypersonic glider is almost unlimited; however, there is no need to exceed approximately 11,000 miles, which would take you from any point on earth to any other point.

The average fare per 1,000 miles has been determined by using airline figures for first-class fares. It costs approximately $115 for each 1,000 miles traveled on the present-day jet transport. Scaling up the costs of equipment, operation, depreciation, and fuel, the supersonic transport fare in 1975 will be approximately $150 per 1,000 miles. The fares for the hypersonic glider in the 1980 to 1990 period were figured on a per-trip basis, giving a range of rates from $178 per 1,000 miles for a 4,000-mile trip down to $77 per 1,000 miles for an 11,000-mile trip. It is recognized that all costs are increasing owing to inflation. This factor has been neglected here since the purpose of the estimate is for comparison only. A trip from Los Angeles to Paris which now costs $710 and requires 10 hours and 45 minutes will, in a hypersonic glider, cost $730 and take only 1 hour and 4 minutes. Interestingly enough, our hypersonic transport would need only 9 minutes longer to go from Los Angeles to Melbourne, Australia, although the distance is nearly 2,400 miles more than the Paris trip (see Table 7).

The route from Los Angeles to Paris extends over central Canada, passes over Greenland, and then crosses England. During the 1-hour transit time, the earth rotation moves the relative position of Paris

TABLE 7. TRANSIT TIMES

(Estimated hours and minutes) *

From Los Angeles to	Distance in miles	Present-day jet	Supersonic jet	Hypersonic glider
Tokyo	5,600	13:12	6:00	1:03
Bombay	8,810	19:37	8:18	1:16
Moscow	6,140	14:17	6:24	1:05
Paris	5,711	10:45	6:04	1:04
Capetown	10,165	24:20	9:14	1:22
Buenos Aires	6,148	12:18	6:23	1:05
Melbourne	8,098	14:12	7:47	1:13

* It is assumed that 2 hours are required for refueling stops when trip distance exceeds the range.

approximately 1,000 miles to the east. This requires that the great circle path be aimed slightly to the east of Paris. This phenomenon, unique to the navigation of hypersonic transports in space, need not be considered for the present-day airliners which operate within the rotating atmosphere. A hypersonic transport traveling from the west to the east benefits from the initial rotational velocity at the earth's surface of approximately 1,000 mph. Hence, it can reduce its top speed by that amount, resulting in fuel economy and lower re-entry temperatures.

Intercontinental terminals of the future will be established on the basis of political and economic activity, and will serve as central locations from which feeder service to the rest of the continents is practical. Remember that the hypersonic transport does not replace the jet transport, except over extremely long-range intercontinental routes. The present-day and supersonic jets will most certainly be required for transcontinental and long-distance feeder service. The jet transports likely will be more practical and economical for freight and mail delivery. However, lightweight and low-volume freight and mail which can benefit from the hypersonic speeds can be included in the payload of the hypersonic transport.

Now, we are off for a trip from Los Angeles to Paris. The time is 10 A.M. in Los Angeles and 7 P.M. in Paris about the year 1990, and the purpose may be something as simple as an international business conference in the Economic Administration Building of NATO.

We have arrived at the new modern airport via a 30-place helicopter. A short walk and an escalator ride brings us to the loading deck for intercontinental flights. A telescoping gangway connects this loading deck to the pressurized doorway of the hypersonic transport. Our walk through the gangway takes us over the wings of an immense air-breathing booster and the wing of the sleek transport (Figure 15).

Upon entering the cabin, we notice a row of small round portholes and a pair of comfortable reclining chairs on each side of a center aisle. Our seats have been previously assigned and we settle down into one of the form-fitting reclining chairs. An attendant will assist us in forming the cushions around the small of the back, the neck, and the head. Then by turning a knob, this molded seat will stiffen in this personally fitted shape. Conventional seat belts and a simple shoulder harness will hold us properly into the molded form. The view from the porthole is rather limited and the glass is several layers thick. It is recessed into the heavily insulated fuselage wall and has a sliding shield between sections of the glass. This closes off the opening from the intense heat which would be radiated through this window during re-entry. On take-off, however, these porthole shields are open.

The pilot and co-pilot have completed their cockpit checks, the airport electronic computer center has determined the flight path and schedule for the trip, and all inputs to the on-board computer have been "dialed in." The attendant straps himself in. Through the individual loudspeakers we are informed by the pilot that the loading has been completed and we are about to taxi to the end of the runway for take-off. There is a series of lights on the seat backs in front of us which will keep us constantly informed on the progress of the flight. The first light to come on illuminates a small sign indicating TAKE OFF. Actually, this take-off is similar to those we have experienced many times in any conventional aircraft. There is a slight sensation of acceleration and, as we look through the window, the airplane is rolling along the runway. The lift-off is so smooth we hardly realize we are in the air. The sound insulation is so complete we can comfortably converse with our seat partner.

Immediately upon take-off, the airplane noses up at a 15° angle and proceeds in a steep climb. As we pass over the 5-mile sound-isolation area, we rapidly gain altitude and, by the time we reach the Pacific shoreline, we are at 10,000 feet. The plane continues to climb in a large arc, attaining supersonic speed somewhere above 50,000 feet. Shortly thereafter, the sound of the engines changes from the muffled high-pitch whine of the turbojets to the low-pitch roar of the ramjets. The window shields are closed by controls in the cockpit.

Approximately 10 minutes from take-off, another indicator light reads SECOND-STAGE BOOST. We are now prepared for the short-duration acceleration flight of the rocket (Figure 17). The sound of the rocket comes from the rear of our transport. We feel an acceleration force as though we were sharply tipping back in our seats and this sensation increases for approximately 4 minutes. It reaches a maximum acceleration, front to back, of three g's, or three times the force of gravity. This force is not unduly uncomfortable and the duration of three g's in this airlaunch, which is less than 1 minute, will not be unbearable for a person in normal health.

The next light to appear reads GLIDING FLIGHT. This, of course, is obvious to us because now the acceleration force has decreased and we have a very lightweight feeling. The acceleration due to gravity is

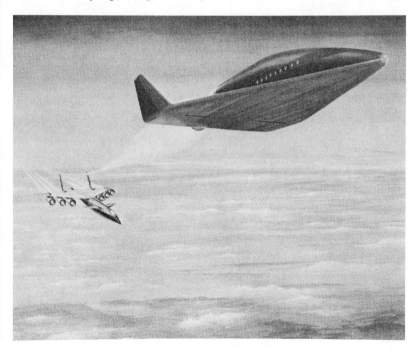

FIGURE 17. Second-stage rocket boost of hypersonic transport.

partly eliminated because of our high velocity and the resultant centrifugal force. We know our velocity now is over 14,000 mph. The outside skin of our transport is beginning to glow red (Figure 18).

The automatic pilot monitored by our captain is gently controlling us through the gliding and re-entry flight. The outside temperatures range from 2500°F at the leading edge to less than 1000° at the trailing edge. The temperature in the cabin remains comfortable, although we can hear the increased air flow through the air-conditioning system. During this period, a SMOKING ALLOWED sign is lighted and smokers have time for about one cigarette before the pilot announces that we are approaching Paris and that our landing is scheduled in 10 minutes. The attendant moves about the cabin and chats with the passengers.

As we enter the lower atmosphere, the acceleration due to gravity has been regained and the flight from here on is the same as any conventional airliner. The porthole shields are again opened and we look out into the evening. It is about 8 P.M. Paris time, 1 hour after take-off.

FIGURE 18. Gliding flight of hypersonic transport.

FIGURE 19. Landing of hypersonic transport.

Guided by an automatic landing system, we enter the landing pattern with the lights of conventional airliners above and below us and a slight noise is heard when our small jet landing engine is started. The touch-down speed is at a smooth 200 mph (Figure 19). We decelerate to normal taxiing speed after a 5,000-foot ground run and taxi up to the terminal building for unloading. The external skin now has cooled to ambient temperature and we leave the transport to join the throngs at the airport, visibly excited with our space-flight adventure.

"Wait until I tell the folks at home about this," is probably the thought uppermost in our minds.

And now, in conclusion, to paraphrase Dr. Cornog's 1956 prediction: I would like to say that while we may have been guilty of errors of concept, in assumptions made, in the numerical calculations, or in the details of conclusions drawn in this chapter, it is our conviction that such manifold transgressions do not invalidate our main conclusion; namely, that the rocket-propelled hypersonic space-glider airplane can be a safe, technically sound, economical form of terrestrial transportation in tomorrow's space age that we clearly see ahead of us.

6

Studying the Universe from a Space Platform

LEO GOLDBERG

HIGGINS PROFESSOR OF ASTRONOMY, HARVARD UNIVERSITY
AND STAFF MEMBER, SMITHSONIAN ASTROPHYSICAL OBSERVATORY
CAMBRIDGE, MASSACHUSETTS

Ph.D. (Astronomy), Harvard University, 1938. Awarded the Bowdoin Prize in 1938 while at Harvard. Dr. Goldberg went to the University of Michigan in 1941 as an assistant at the McMath-Hulbert Observatory, followed by research associate in engineering research and research physicist. In 1945 he became Assistant Professor of Astronomy and from 1948 to 1960 held the rank of Professor of Astronomy at Michigan. In 1960, he assumed his present position as Higgins Professor of Astronomy, Harvard University, and Staff Member, Smithsonian Astrophysical Observatory, Cambridge, Massachusetts. During World War II, he served as consultant to the U.S. Navy, Bureau of Ordnance, and received an Individual Award for Exceptional Service to Naval Ordnance Development in 1946. He is Vice President of the American Astronomical Society, Vice President of the International Astronomical Union, and member of other professional organizations. He is a Fellow of the American Academy of Arts and Sciences and a member of the Board of Trustees of the Cranbrook Institute of Science. He is also Chairman, U.S. National Committee of the International Astronomical Union. He is also a member of the National Academy of Sciences, American Philosophical Society, Space Science Board of the National Academy of Sciences, Board of Trustees of Associated Universities, Inc., Chairman of the Astronomy Committee of the Space Science Board, and is a Foreign Associate of the Royal Astronomical Society of Great Britain. His chief specialty is solar physics, and he has recently undertaken work leading to the construction of astronomical instrumentation for satellite observatories.

IN THE long history of astronomy, there have been many great discoveries and break-throughs that have carried man past one frontier after another in his search into the nature of the universe. One of the most spectacular of these occurred on October 10, 1946, when a camera in the nose of a V-2 rocket photographed the sun's ultraviolet radiation from a height of nearly 100 kilometers above the New Mexico desert. The existence of this radiation had been inferred for a long time, but it was also known that ultraviolet rays cannot penetrate through air, and very few astronomers had even dared to hope that within their lifetime instruments could be transported to the great altitudes necessary to get above the earth's atmosphere. The success of this feat by Dr. Richard Tousey and his associates of the Naval Research Laboratory in Washington shows how quickly the nearly impossible can become reality in the present explosive atmosphere of science and technology. The pace is so rapid that it sometimes leaves even the best astronomers breathless. After seeing the first rocket photographs of the sun, the late Prof. Henry Norris Russell of Princeton University wrote as follows: "These rocket spectra are certainly fascinating. My first look at one gives me a sense that I was seeing something that no astronomer could expect to see unless he was good and went to heaven!"

Since 1946, dozens of astronomical rockets have been flown to higher and higher altitudes with better and better instruments. For a few fleeting moments during each flight they have given us glimpses of the universe which are radically different from those seen by ground-based telescopes. They are like scenes shown in a movie theater to advertise a coming attraction, the full drama of which will unfold when astronomical telescopes are placed in a more or less permanent satellite of the earth or in an observing station on the moon.

Astronomers have been working under very severe handicaps in their efforts to understand the universe. To begin with, almost everything that has been learned about the universe has come from observation at very great distances, except when occasional chunks of matter in the form of meteorites have fallen upon the earth and have been studied in the laboratory. But the astronomer may now look forward to probing at least the nearby environment of the solar system with vehicles carrying packages of instruments that can make measurements and observations and relay the information back to earth by radio. It even seems altogether probable that in the not too distant future samples of the surfaces of the moon and some of the planets will be returned to earth for examination. The astronomer also now has it within his power to create artificial satellites and planets and to place them at locations and in orbits of his own choice. So we see that the astronomer will soon become an explorer and an experimenter as well as an observer.

But the universe is so vast that even if our capability for sending larger and larger vehicles to greater and greater distances should exceed our wildest dreams, astronomers will still have to rely on observation to investigate the universe beyond the solar system. However, a tremendous advantage will be gained when the astronomer's observing station is moved just a few hundred miles from the surface of the earth, above the muddy and turbulent ocean of air that obscures his view from the ground. The air is a three-way nuisance to astronomy. First, the atoms and molecules that make up the atmosphere are like so many tiny reflectors or mirrors surrounding the earth. During the day they intercept the rays of the bright sun and reflect or scatter them in all directions—most aggravatingly in the direction of the earth. This is why the daylight sky is bright and stars are invisible in the daytime. Even at night, the sky is never dark but glows with the light of the permanent aurora. The bright glow of the aurora borealis is a familiar sight to those who dwell in northern latitudes. The brilliant apparitions shown in Figure 20 are seen rather infrequently and for short periods of time, except in the far north. They are caused by the bombardment of the high atmosphere by fast-moving electrified particles emanating from the sun. A bright aurora is usually seen only after a storm has occurred on the sun. At other times, smaller numbers of particles more or less continuously stream into the atmosphere causing its atoms to emit a faint glow known as the permanent aurora. This glow is not strong enough to be visible to the naked eye, but is nevertheless bright enough to mask the faint light from distant stars and galaxies, the detection and measurement of which is of great importance for stellar evolution and cosmology.

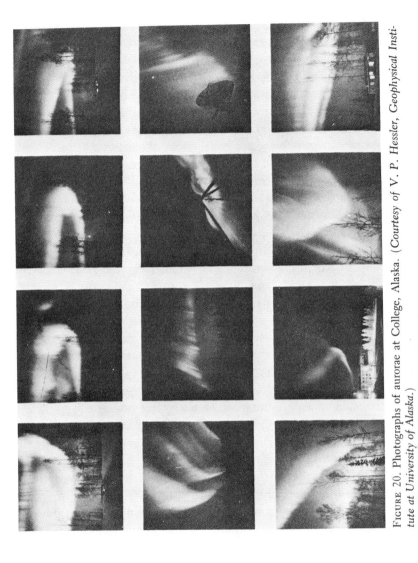

FIGURE 20. Photographs of aurorae at College, Alaska. (Courtesy of V. P. Hessler, Geophysical Institute at University of Alaska.)

The second property of the earth's atmosphere that is troublesome to astronomers is its turbulence. The atmosphere is in a constant state of agitation and so smears out fine details in the forms of the sun, the moon, the planets and the nebulae. The astronomer calls this condition "bad seeing." The degree of turbulence varies from place to place and also changes with time, but it is always the factor that limits the quality of observations made from the ground. Even under the best possible seeing conditions on Mt. Palomar, star images photographed with the 200-inch telescope are ten times as large as they would be in the absence of an atmosphere.

How high up in the atmosphere must one go before bad seeing is entirely eliminated? This question has been answered in brilliant fashion by Prof. Martin Schwarzschild, an astronomer from Princeton University, who has succeeded in obtaining crystal-clear photographs of the sun from an unmanned balloon that carried a telescope and a camera 80,000 feet into the stratosphere. One of his photographs is shown in Figure 21. Observations from the ground had revealed that the surface of the sun is not smooth and featureless, but that it appears to be broken up in a granular pattern. These granules have been a subject of great interest to astronomers for more than a century. More recently, it has become clear that they are the consequence of a huge circulation system in the sun's atmosphere. The bright granules are the tops of columns of hot gas—bright because they are hot—rising upwards, whereas the darker regions contain material descending after it has cooled. But the precise character of the convection has eluded description because bad seeing made it impossible to determine the exact sizes and shapes of the granules, information which can now be obtained from the balloon photographs.

We come next to the third and most important reason for placing astronomical telescopes above the earth's atmosphere, which is that the air is not transparent and is, in fact, about as effective as a brick wall in obstructing man's view of the heavens. At first this sounds unbelievable because everyone knows that there are days and nights when the sun and the stars shine brightly through what appears to be a perfectly transparent atmosphere. The answer is that although the atmosphere is indeed transparent to visible light (which is the only kind of radiation that the eye can perceive), visible light is only one of many forms of radiation.

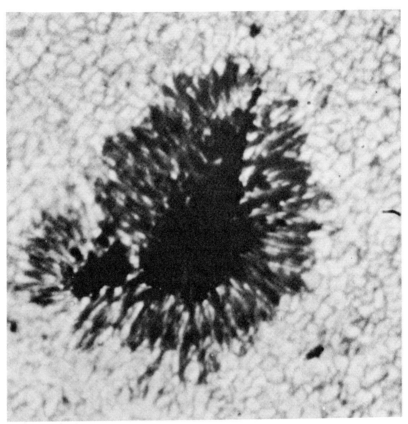

FIGURE 21. A large sunspot on the sun. Photographed from an unmanned balloon at a height of 80,000 feet on August 17, 1959. (*Courtesy of M. Schwarzschild, Princeton University Observatory.*)

In addition to visible light, a ball of hot gas like the sun or any other star radiates gamma rays, X rays, ultraviolet, infrared, and radio waves. But only the visible and some parts of the ultraviolet, infrared, and radio regions of the spectrum pass through the atmosphere unscathed, as shown in Figure 22. Ozone, nitrogen, and oxygen are powerful absorbers of far ultraviolet and X radiation and not even a glimmer gets through the atmosphere. A certain amount of infrared radiation does penetrate to the ground, but most of it also is swallowed up in the atmosphere by such constituents as water vapor and carbon dioxide. Short radio waves can also get through but those longer than about 30 meters are reflected back into space by the electrified layers of the earth's upper atmosphere. So effective is this atmospheric screen that it can only be avoided entirely from an altitude of about 150 miles, where the density is much less than one-billionth of its value at sea level.

You can readily see that the astronomer has been in the position of a reader trying to unravel the plot of a mystery novel from which most of the pages have been torn.

I should like to turn now to a brief survey of some of the important unsolved problems that are currently occupying the attention of astronomers and to indicate how observations from space vehicles can aid in bringing about their solution.

Quite apart from its close proximity to the earth, the moon (Figure 23) is an object of very great interest to astronomers because its study can tell us a great deal about the origin and evolution of the entire solar system. Because of its nearness to the earth, it is very doubtful that any radically new information can be had by observing the moon from a balloon or even from a satellite of the earth. The questions that we ask about the moon can probably not be answered without sending instruments close to or upon its surface: What is the exact shape of the moon and how does its density vary throughout its interior? Does the moon have a magnetic field? What is the composition of its surface and does it vary from one part to another, as, for example, between the mountains and the seas? Does the moon have even a trace of an atmosphere? If it does, it would be an exceedingly thin one containing perhaps as little as one-billionth as much gas as the earth's. Gravity on the moon is very weak and therefore any light gases would escape com-

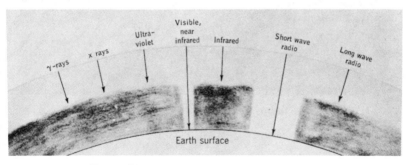

FIGURE 22. Transmission of earth's atmosphere.

FIGURE 23. The moon, age 22 days. (*Courtesy of Lick Observatory.*)

pletely, but heavy ones like carbon dioxide and krypton might be present in very small quantities.

Many of these and other questions about the moon could be answered by flying an instrumented satellite close to its surface; others would require the landing of instrument packages and radio transmitters. Eventually it will be necessary to return samples of lunar material for analysis in terrestrial laboratories, and it is possible that apparatus for this purpose can be developed in the future. It may be, however, that a manned landing on the moon will be required before complete answers to its nature and origin can be given.

Aside from the moon, the exploration of other planets in the solar system is bound to be a primary goal of man's venture into space. One very compelling incentive is the sheer urge of the human race to explore the unknown. The scientific motivation is also very strong, since knowledge of the nature and composition of the planets is almost certain to tell us how the solar system was formed. Even more exciting would be the discovery of life on Mars or Venus. We are now fairly sure that some form of plant life exists on Mars, as evidenced by the dark greenish colorations that appear during the Martian spring. The possibility of studying this and other forms of life that may be discovered is a fascinating prospect for the human race.

The distortions caused by the earth's atmosphere make it most probable that very little further will be learned about the planets by observations from the ground. Balloons and earth satellites will greatly advance our knowledge, but the most significant revelations will be produced by packages of instruments sent out in space probes. As time goes by, the instruments will be flown ever closer to the planets. Eventually, they will be made to orbit about the planets, and finally, they will be lowered to the surfaces. The observations and measurements that will be made will be of many kinds, including television of planetary landscapes, studies of magnetic fields, determinations of the temperatures and compositions of the atmospheres and surfaces, experiments for the detection and study of biological life-forms, and many others.

We next turn our attention to the sun (Figure 24), which is both the nearest star and, for the human race, the most important one. It is basically a globular ball of gas, composed principally of hydrogen and helium with a sprinkling, perhaps a per cent or two, of the other, heavier elements. Its average density is a little greater than that of water, but

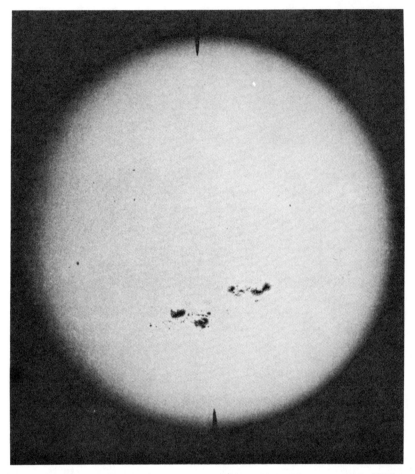

FIGURE 24. The sun, showing large sunspots, July 31, 1949. (*Courtesy of Mt. Wilson and Palomar Observatories.*)

the bulk of its material is concentrated chiefly near the center where the density is about seventy times greater than the average value. Needless to say, the density of its outer parts is exceedingly minute, being comparable to that which would be found in a good vacuum in an earthly laboratory. The sun's temperature also decreases steeply from the center outwards to the visible surface. It is about 15,000,000°C at the very center, where mass is converted into energy by the fusion of hydrogen nuclei into helium. As a result, the sun loses mass at the rate of about 4,000,000 tons each second. So huge is the sun, however, that even at this rate it casts off only 1 per cent of its mass in a little over 1 billion years. The visible surface of the sun, where the temperature has fallen to about 6000°C, has a diameter of 864,000 miles.

The sun also has an outer envelope that consists of two parts, a relatively narrow zone about 10,000 miles thick called the *chromosphere*, and above it, an enormously distended halo, the *corona* (Figure 25), which reaches out for tens and perhaps even hundreds of millions of miles into the solar system. So weakly does the outer envelope radiate, it cannot be seen against the bright daylight sky, but it can be seen in all its splendor during a total eclipse. Most of the drama and mystery

FIGURE 25. The solar corona, photographed by G. van Biesbroeck during the total solar eclipse of February 25, 1952, at Khartoum. (*Courtesy of National Geographic Society.*)

connected with the sun are associated with events that take place in
these outer layers. For example, up until 20 years ago, it was taken for
granted that the sun became cooler with increasing distance from its
center. But in 1941 the Swedish physicist Edlén made the startling discovery that although the sun's temperature did indeed decline steadily
from the center outwards to the visible surface, the trend was sharply
reversed in the chromosphere where the temperature rises abruptly,
and in the space of a few thousand miles attains a value of about
2,000,000° in the corona. Even at the distance of the earth the gas may
still be as hot as three or four hundred thousand degrees. But our
knowledge of the chromosphere and corona is still very meager; first,
because most of this knowledge has been gained during the few fleeting moments of solar eclipses and, second, because the temperatures in
these regions are so high that their radiation consists largely of ultraviolet and X rays, which up until now could only be studied during
rocket flights lasting but a few minutes.

The atmosphere of the sun, like that of the earth, is often the scene
of short-lived and violent activity, but on an incredibly larger scale.
These disturbances usually break out close to a sunspot, and, indeed,
both the frequency with which they occur and the size of the event go
roughly up and down with the sunspot cycle, the 11-year period during
which the number of spots on the face of the sun waxes and wanes. The
sunspots themselves are located rather deep in the sun's atmosphere,
but the disturbances associated with them occur much higher up, in its
chromosphere and corona. By all odds, the most catastrophic of these
disturbances is the solar flare, in which a stupendous amount of energy
is generated and released in a relatively small region during periods of
time varying from a few minutes to a few hours. One of the largest
flares ever observed on the sun, on July 16, 1959, is shown in Figure 26.
The total amount of energy let loose in this way can amount to more
than one billion billion kilowatt hours. The onset of a flare is most
commonly advertised by a sudden brightening of the visible light from
the affected region and by great bursts of radio waves, often at all
frequencies in the radio spectrum, but most commonly at the lower
frequencies in the band from 600 to a few megacycles. It is now also
known from rocket observations by the Naval Research Laboratory
that flares also emit bursts of X rays. Thus solar flares radiate intensely
from one end of the electromagnetic spectrum to the other, and since

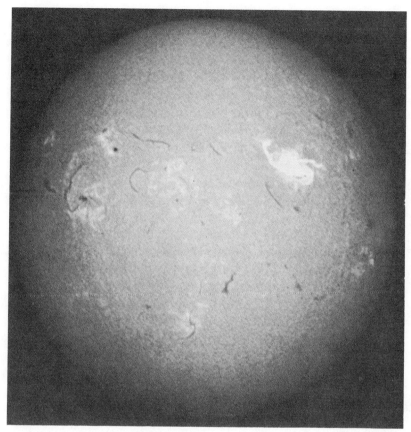

FIGURE 26. The great solar flare of July 16, 1959. (*Courtesy of McMath-Hulbert Observatory, University of Michigan.*)

only a small fraction of this radiation reaches the surface of the earth, it is small wonder that the mechanism of origin of a flare is so poorly understood.

Not all of the energy released by a flare is in the form of radiation. Great streams of electrified particles are also hurled into interplanetary space at speeds ranging from a few hundred miles per second up to some tens of thousands of miles per second in the case of the brightest flares. Even when the sun is quiet, a certain number of these particles are blown out in a more or less steady stream in all directions through the solar system. This phenomenon is in effect a solar wind, which can attain hurricane force at times of solar flares.

The impact of a solar flare upon the earth is often very dramatic; some of its effects are visible almost instantaneously while others occur after a number of hours or even days. For example, the X rays travel away from the sun at the speed of light and therefore reach the earth in about 8 minutes. The earth's atmosphere protects us from the onslaught but only after the atmospheric atoms and molecules have suffered the loss of large numbers of electrons ripped off by the incoming X rays. As a result of the increased electrification of the atmosphere, radio waves broadcast by a transmitter on the earth are no longer reflected back by the ionosphere and thus enabled to reach receivers at distant points. Instead, the radio waves are absorbed and hence there is an abrupt cessation of long-distance radio communications which may endure for several hours.

Solar flares also cause the earth to be bombarded by a rain of high-speed electrified particles. The most energetic particles travel at speeds which are an appreciable fraction of the speed of light and may reach the earth after less than an hour. They carry enough momentum to crash through the barrier of the earth's magnetic field and strike the ground, where they are recorded as cosmic rays. The slower-moving particles do not get through the magnetic barrier, but they do disturb the equilibrium of the inner Van Allen radiation belt, which is thereby caused to unload a portion of its store of charged particles into the atmosphere at the higher latitudes in the neighborhood of 70°. The bombardment of atoms in the high atmosphere by these entering charged particles causes brilliant displays of northern lights. Furthermore, the motion of the charged particles creates electric currents which modify the earth's magnetism in an important way, causing compass needles to fluctuate wildly and long-distance telephone and telegraph communications to be savagely disrupted.

It will be readily apparent that events on the sun profoundly affect the course of life on earth and therefore that the nature of these events and the circumstances of their origin must be better understood if we are to learn to live with them. Were it not for the earth's atmosphere, the highly lethal and poisonous radiations from the sun would make life on earth completely impossible. However, at the same time that these harmful radiations are shielded by the atmosphere, they are also hidden from the view of the astronomer who is thereby prevented from discovering the true nature of the sun. The sun is near enough so that it is within striking distance of the instrumented vehicles that will be sent out to roam about the solar system within the next few years. Already the probe Pioneer V has sent back accounts by radio of streams of solar particles traveling through space many millions of miles from the earth, including a report from a point 13 million miles nearer to the sun than the distance of the earth. The sun was still 80 million miles away, but much closer approaches, perhaps even to within a few million miles, might be feasible before a vehicle and its instruments would be melted and vaporized by the intense heat. Solar probes will be invaluable for the study of solar winds, but in the final analysis these winds are controlled by events on the sun itself, which must be studied on a continuous 24-hour basis from observational platforms provided by artificial satellites.

The sun is a relatively easy object to observe; its great apparent size and brightness enormously simplify the problem of finding it in the sky from a satellite and keeping a telescope locked on. Observing the stars from a platform in space will be many times more difficult, but there are no inherent limitations that we can now foresee and certainly the scientific rewards will justify whatever effort needs to be made. Probably the most important discoveries that will come from the study of the stellar universe will be totally unexpected, but nevertheless astronomers have been studying the universe for a long time and have developed numerous theories and hypotheses which can be subjected to crucial tests from the vantage point of an observatory in space. By all odds the most important of the theories is concerned with the past history of the universe and with such questions as "When and how were the stars and galaxies formed?" and "In what manner did they evolve into their present forms?" These are enormously difficult questions, which are not likely to be answered in our lifetime, but it will not be because we have not tried. Let me try to summarize some of our present notions concerning the origin and evolution of stars and galaxies and then to suggest how these ideas may be reinforced, modified, or rejected by observations in space.

The sun is one of a galaxy of stars, perhaps one or two hundred billion in number, which are grouped together in the form of a highly flattened disk about 100,000 light-years in diameter, one light-year being about 6 million million miles. The observable universe appears to be filled with roughly similar galaxies distributed throughout space

LEO GOLDBERG

at least as far as the great 200-inch telescope can penetrate, which is about 2 to 3 billion light-years. The exact structure of our galaxy is not known since we are on the inside looking out, so to speak, but we know that it cannot be greatly different from one of its near neighbors in space, the Andromeda Galaxy (Figure 27), which is about 2 million

FIGURE 27. The Andromeda spiral galaxy. (*Courtesy of Mt. Wilson and Palomar Observatories.*)

light-years away. We observe that in this galaxy, as well as in our own, the space between the stars is filled with great clouds of gas and dust (Figure 28); furthermore, the brightest stars, together with the gas and dust, tend to be distributed along spiral arms which open out from opposite sides of the central nucleus. There is good reason to believe that about 10 billion years ago our galaxy consisted entirely of gas and dust from which the stars were later formed. The primordial gas was almost pure hydrogen, which is the lightest of all the elements, and the building block from which the heavier elements are constructed by nuclear fusion. The fusion of elements is a by-product of the process by which the energy that heats a star is generated in its interior. The life history of a star is intimately connected with the process of energy generation, and indeed the length of its life is determined precisely by the rate at which it consumes its fuel.

Near the beginning of its life, the star begins to convert its hydrogen into helium, but after a certain fraction of the hydrogen is consumed, the formation of helium ceases, the star undergoes rather radical changes

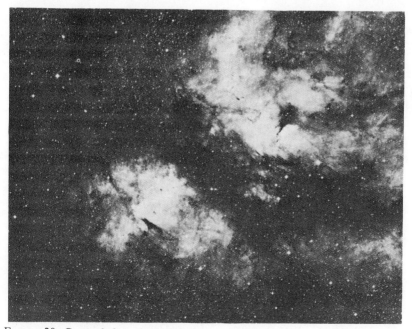

Figure 28. Gas and dust clouds near the star Gamma Cygni. (*Courtesy of Mt. Wilson and Palomar Observatories.*)

in size and structure, and it is transformed from a normal star into a so-called red giant. Later, the generation of energy is resumed but this time with helium as the fuel and still heavier elements as the end product. This process goes on and on until eventually all sources of nuclear energy are used up. In the latter stages of its life the star almost literally explodes, ejecting its outer layers into interstellar space, and leaving only a small dense core of material in the form of a feebly glowing white dwarf star. This account of the life history of a star is far from complete, but it is a good working hypothesis that seems to explain at least the gross features of the observations.

We see therefore that hydrogen is the raw fuel of stellar energy and that the heavier elements are the ashes. Thus, when a dying star explodes its material into space, the interstellar gas is enriched by the addition of a sprinkling of heavy elements and therefore its original composition is modified. How long does this evolutionary process take? In other words, what is the age of a typical star? The answer is that the age depends on the rate of consumption of its hydrogen fuel, and this in turn depends upon the mass of the primeval star. It is an observational fact that the most massive stars are also the most luminous and therefore they must consume their fuel much more rapidly than less luminous stars. Consequently their lives will be shorter. We know of some superluminous stars that are using up energy at such a prodigal rate that they cannot possibly have been shining for more than a few million years. On the other hand, the majority of stars shine so feebly that they can last for tens and even hundreds of billions of years.

The picture we have is one in which the stellar population of our galaxy includes all ages from a few million to a few billion years. Indeed, like any population, some of its members at this very moment are dying while others are in the process of being born from the interstellar gas and dust. If this hypothesis is correct, there should be significant differences in chemical composition between the old stars and the new. The youngest stars have been created from material which has been enriched or contaminated by the debris from dying stars, and therefore they should contain a higher percentage of heavy elements relative to hydrogen than do the older or first-generation stars. Thus, if the chemical compositions both of stars of many ages and of the interstellar gas could be determined with precision, it would provide the most sensitive kind of test of our present ideas on the history of the universe.

This brings me to one of the major reasons why astronomers wish to establish an observatory in space. There is of course no possibility that we can ever scoop up a sample of a star and bring it home for chemical analysis. But the chemical content of any gaseous mixture of elements can be ascertained no matter how far away it is, as long as it emits or absorbs radiation which can be received and analyzed by a spectroscope. The use of this technique by astronomers over many decades has established the presence of the familiar chemical elements everywhere in the universe and in roughly the same proportions as they are found on earth. But the measurements have been made only with visible radiation and they have not been sufficiently precise to provide the answer to the puzzle of stellar evolution. There are good reasons for believing that the ultraviolet radiation is a much more sensitive indicator of chemical abundances. Thus, the stake of astronomy in space is no less than the comprehension of how and when the universe came into being, of how it has progressed to its present state, and of what its future is likely to be in the billions of years that lie ahead.

The detailed research programs that must be carried out before the goals of astronomy in space can be realized are now being worked out in the minds of many astronomers in many countries. In the United States, the broad outlines of a national space program for astronomy have begun to take shape and some of the details are starting to be filled in. The blueprints for the program cannot be drawn precisely because space technology is in such a state of rapid flux that one cannot be certain too far in advance of the rocket and instrument capabilities that will be available for use a few years from now.

It is not unreasonable to assume, however, that it will be possible within the next 20 years to establish a very large astronomical observatory in space, either in an orbit around the earth or on the moon. By "large" I mean one that contains equipment and apparatus measured in some tens of tons. The observatory would contain a variety of telescopes up to perhaps 100 inches in diameter, as well as one or more radio telescopes and the usual accessories in the form of photometers, spectroscopes, image tubes, and radio astronomy receivers.

At first glance, it might seem that the moon would be the preferred location for such an observatory because it seems to offer a more stable foundation or platform than would a vehicle in orbit. Actually, it is perfectly feasible to launch a large telescope into orbit around the earth, to point it accurately at any chosen star by radio command from the earth, and to receive data at a ground station, also by radio. The absence of gravity in a satellite is a very great advantage in achieving accurate pointing, because even on the moon the force of gravity is still one-sixth as great as it is on the earth. Furthermore, the closeness of the satellite to the earth makes it much easier to maintain radio communication and control than would be the case on the moon. It has also been argued that an astronomer could function better on the moon than in a space station, but it is also not clear that the physical presence of man will even be necessary for the successful operation of the observatory. As far as a satellite observatory is concerned, the presence of a man in the same package with the telescope would be a serious and almost impossible liability to overcome, because even the slightest movement or shifting of weight would cause the satellite and telescope to revolve and would make accurate pointing well nigh impossible. Thus, the role of man in the operation of a space observatory is likely to be limited to occasional visits to replace worn out parts or to change equipment attached to the telescope. How frequently such visits would have to be made would depend upon the reliability of the equipment. From this point of view, the moon does offer an important advantage over a space station, because out in space man would be exposed to bursts of highly lethal cosmic radiation from solar flares from which he could escape by going underground on the moon. The moon also offers important advantages for radio astronomy because it could provide a rigid foundation for very large antennas as much as a few miles long. Furthermore, man-made interference caused by radio transmitters on the earth and in satellites constitutes the most serious kind of threat to radio astronomy from which it may only be possible to escape by putting radio astronomy observatories on the other side of the moon.

Many unanswered questions will have to be cleared up before a final decision can be made on the location of a large space observatory, but in the meantime plans are going ahead for the launching of a number of astronomical satellites containing relatively modest equipment. I shall choose a few examples to illustrate the general size and shape of the United States program for the next 5 to 10 years.

To begin with, instrumented rockets will continue to play an important part in the space program, in the testing of new ideas for equipment and in making preliminary surveys of radiation from the sun and stars. The Aerobee-Hi rocket has been used with great success for these purposes. As shown in Figure 29, it is equipped with a pointing control which keeps the instruments in the nose cone locked on the sun during flight.

The first satellite specifically designed for astronomical research will be launched in 1961 by a Thor-Delta rocket. The satellite is now under design and construction by the Ball Brothers Research Corporation for the National Aeronautics and Space Administration and will be known as an orbiting solar observatory. As shown in Figure 30, it will have the shape of a wheel about 4 feet in diameter and will weigh about 75 pounds. A fan-shaped array of solar battery cells will supply power for the operation of the equipment and for the transmission of data by radio to the earth. The vehicle will be stabilized by gas jets so that the instruments housed in the two boxes will point accurately to the sun at all times. The three balls contain a supply of gas for the jets and the arms holding them also serve as radio antennas. The orbiting solar observatory will carry ultraviolet spectroscopes, X-ray and gamma-ray telescopes, as well as radio astronomy antennas and receivers, which are being designed and constructed by a number of groups at the University of Colorado, Harvard University, the University of Michigan, and the Goddard Space Flight Center of NASA.

FIGURE 29. Nose cone of Aerobee-Hi rocket showing biaxial pointing control. (*Courtesy of Ball Brothers Research Corp.*)

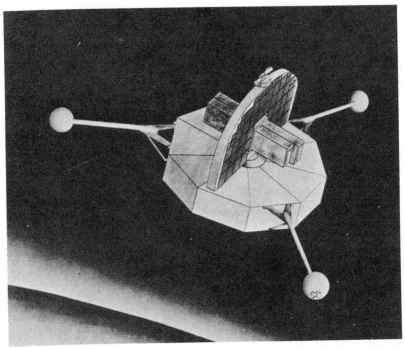

FIGURE 30. Orbiting solar observatory. (*Courtesy of Ball Brothers Research Corp.*)

Sometime later, perhaps in 3 or 4 years, a much larger and more sophisticated observatory will be launched into orbit by an Atlas rocket. It will go by the name of Orbiting Astronomical Observatory, will weigh several tons, and will permit highly accurate pointing and tracking by telescopes to a precision of a fraction of a second of arc.

A number of different universities and government laboratories are preparing astronomical equipment to go on board this satellite, including the Smithsonian Astrophysical Observatory, Princeton University, the University of Wisconsin, Harvard University, and the Goddard Space Flight Center. The program will include mapping of the entire sky in several ultraviolet colors, the analysis of the chemical composition of the interstellar gas, the observation of ultraviolet radiation from individual stars, and the detailed examination of the sun in ultraviolet and X radiation. The engineering problems that must be overcome before the goals of the program are achieved are fearsome in their complexity. During launching, the delicate and fragile optical and electronic equipment will be subjected to accelerating forces of about twenty times gravity and to violent shaking forces of about the same amount. Once the equipment is in orbit, it will be continuously bombarded by ultraviolet and X rays, by high-speed electrified particles, and by small meteors, all of which can cause damage of one kind or another. During half of its orbit, the satellite will warm up in the hot rays of the sun, and during the other half will turn frigidly cold. The satellite telescope will be expected to point at individual stars with fantastically high precision for long periods of time despite the many disturbing forces that will be at work to cause the satellite to rotate and tumble. Among these are the pressures exerted by radiation from the sun and the earth, forces akin to the tides caused by the earth's gravity, and the disturbing effect of the earth's magnetism. The last force is especially troublesome because of changes in the earth's magnetism induced by storms on the sun.

Despite the severity of the engineering problems, those concerned with the program are convinced that they can be solved and that telescopes as large as 2 to 3 feet in diameter can be put into orbit and operated successfully and efficiently within the next few years. A conceptual design of such a telescope has already been carried out under the direction of Dr. Lyman Spitzer of Princeton University as shown in Figure 31. This telescope is designed primarily for the investigation of interstellar gas clouds. It is basically a reflecting telescope, 24 inches

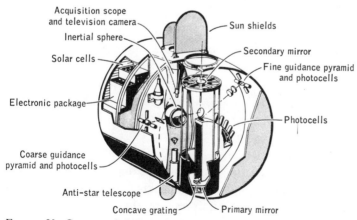

FIGURE 31. Conceptual design of telescope for orbiting astronomical observatory. (*Courtesy of L. Spitzer, Princeton University Observatory.*)

in diameter, to which is attached a spectroscope and a battery of photo-cells for the analysis of ultraviolet starlight. In addition to the telescope and spectroscope, the instrument package contains necessary electronics, solar battery cells, devices for coarse and fine guidance, and a television camera that will transmit a picture of the sky to the observer on the ground, who will thereby be enabled to select for study individual stars of his choice and by radio commands to keep the telescope accurately pointed to the star.

The satellites I have described so far will orbit at an altitude of about 300 miles from the surface of the earth. The advantage of such close satellites is that it is easy to keep in touch with them by radio and, of course, much greater rocket power would be required for a very distant orbit. On the other hand, a far-away orbit also has its attractive features. For example, if a satellite is placed into orbit at a height of 22,500 miles, it will turn about the earth once every 24 hours and thus appear to hover above a fixed point on the earth. If a ground-controlled station is established at this point, it will always be in direct communication with the satellite, and the relatively slow rate of movement of the satellite with respect to the stars would simplify the problem of pointing a telescope. This concept of a 24-hour orbit has been advocated most strongly by A. B. Meinel of the Kitt Peak National Observatory at Tucson, where the National Science Foundation is sponsoring the development of a satellite telescope more than 4 feet in diameter. A preliminary concept of such a telescope is shown in Figure 32. Unlike the orbiting observatory, in which a telescope would be mounted on a stabilized platform, Meinel visualizes an orbiting telescope with built-in devices for guidance and stabilization. This telescope is many years in the future because a total rocket thrust of more than 1 million pounds would be required to boost it into an orbit 22,500 miles in height. But the job could be done by the Saturn rocket now under development. Figure 33 shows how the orbiting telescope would be mounted in the nose of the Saturn rocket at launching, and Figure 34 as it would appear in orbit.

This has been a brief and admittedly incomplete survey of the plans that United States astronomers are making to project their observatories into space. The size and scope of these plans make it certain that the scale on which astronomical research is conducted will be enormously expanded and, I may add, it will be very costly. For ex-

FIGURE 32. Preliminary concept of 50-inch space telescope of Kitt Peak National Observatory. (*Photo by Ray Manley.*)

ample, a single Atlas-launched Orbiting Astronomical Observatory will cost about as much as a 200-inch telescope in both money and manpower, and we may properly ask ourselves whether the satellite astronomy program is not too ambitious and costly, or whether the money might not be more usefully spent for astronomy by building more 200-inch telescopes. The fear has also been expressed that support for ground-

FIGURE 33. Model of space telescope mounted in nose of Saturn rocket. (*Photo by Ray Manley.*)

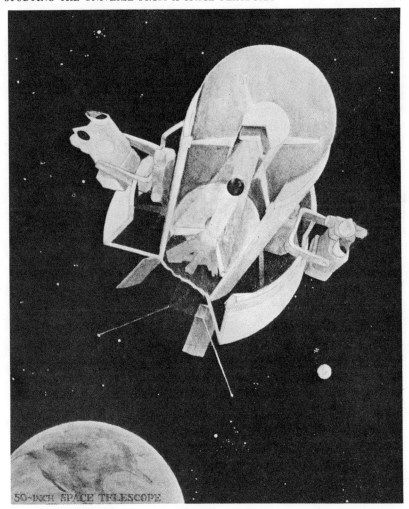

FIGURE 34. Artist's conception of space telescope in orbit. (*Courtesy of Kitt Peak National Observatory.*)

based astronomy will be diverted to the more glamorous projects in space.

I do not share these misgivings for two reasons. First, the public attention that has been focused on astronomy by space vehicles must inevitably lead to increased support for all types of astronomical research, both from the ground and from satellites. Second, it is fruitless and irrelevant to attempt to equate the cost of a satellite telescope to that of a 200-inch telescope, as long as a satellite telescope only a few inches in diameter can get information about the universe that could not be acquired by any number of 200-inch telescopes on the ground. I do not suggest that the 200-inch has become less valuable because of the new developments, but it is no more competitive with a satellite telescope than is an ocean liner with an airplane. This does not mean that astronomers do not have a heavy responsibility to use limited funds and manpower wisely and efficiently. There is a danger that we may waste our efforts on space experiments that can be done better and more cheaply on the ground or that will yield only trivial results. I am personally confident that this will not happen and that the collective judgment of the scientific community will make the fullest use of the golden opportunity that is now available to them.

7

Part I. The Sun and the Earth

JOSEPH KAPLAN

PROFESSOR OF PHYSICS, UNIVERSITY OF CALIFORNIA, LOS ANGELES
AND CHAIRMAN, U.S. NATIONAL COMMITTEE FOR
INTERNATIONAL GEOPHYSICAL YEAR

B.Sc., Johns Hopkins, 1924; Ph.D. (Physics), 1927. National Research Fellow, Princeton University, 1927 to 1928. University of California, Los Angeles, 1928 to date. Chairman of the Meteorology Department, 1940 to 1943. Chairman of the Physics Department, 1945 to 1950. Specialties: molecular spectroscopy and composition of the upper atmosphere. Chief of Operations Analysis Section, 2nd Air Force, and later Air Weather Service, 1943 to 1945. Member, Air Force Scientific Advisory Board since 1947. Member of numerous national committees relating to geophysics, geodesy, astronomy, meteorology, geomagnetism in the upper atmosphere. Chairman, U.S. National Committee for International Geophysical Year. He has received honorary D.Sc. degrees (1956) from University of Notre Dame and Carleton College.

I N 1956—only 4 years ago—a distinguished student of the sun made the following remarks: "It is an unhappy fact of the present state of our knowledge of solar physics that our knowledge of the variable components of solar radiation which produce terrestrial effects depends almost exclusively on the existence and nature of the terrestrial effects themselves. In general we do not observe the emission of far ultraviolet radiation and corpuscles from the sun directly, although we know the emission occurs because we observe the terrestrial reactions. I do not mean to minimize here the brilliant rocket observation of the far ultraviolet region of the spectrum by the Naval Research Laboratory and the University of Colorado. They confirm deductions both from astrophysical and geophysical theory, but observations of this type are still not on a routine basis, and were made for purposes of confirmation." He then expressed the hope that the task of understanding and using the solar terrestrial effects will be made far easier by continuous observations from above the atmosphere.

The dramatic fact is that in less than 4 years since this able and relatively young man made such a statement, observations from satellites and space probes have almost become routine; the story of the sun's ultraviolet, corpuscular, and X radiations is being revealed to us at a rate that will challenge our energies and minds to absorb and use.

I've been greatly impressed in recent times by the notable role that the sun has played in the history of mankind's efforts to understand himself and his universe. Since it is not the purpose of this lecture to recount this fascinating story, I will try to be brief, and here I quote again.

> The Scientific Revolution can be held to begin in the year 1543 when there was brought to Copernicus, perhaps on his deathbed, the first printed copy of the book he had written about a dozen years earlier. The thesis of this book is that the earth moves round the sun.

140

Less than a hundred years after Copernicus, Kepler published (between 1609 and 1619) the three laws which describe the paths of the planets. The work of Newton and with it most of our mechanics spring from these laws. They have a solid, matter-of-fact sound. For example, Kepler says that if one squares the year of a planet, one gets a number which is proportional to the cube of its average distance from the sun. Does any one think that such a law is found by taking enough readings and then squaring and cubing everything in sight? If he does then, as a scientist, he is doomed to a wasted life; he has as little prospect of making a scientific discovery as an electronic brain has.

I quoted here from the book *Science and Human Values* by J. Bronowski, because of the simple way in which he characterizes the way in which Copernicus thought. He did not find his answer by routine calculations, but by a leap of imagination—by lifting himself from the earth, and placing himself wildly, speculatively into the sun. "The earth conceives from the sun," he wrote, and "the sun rules the family of stars."

It is tempting to say that the start of another scientific revolution, perhaps comparable with the one that is identified with the name of Copernicus, occurred when the first earth satellite was launched during the International Geophysical Year. And it is equally tempting to emphasize the role played by the sun in all of this. Perhaps by the end of this chapter you will be convinced that such a statement is not an overly extravagant one.

From the very start of its conception, the IGY was planned to coincide with a period of enhanced solar activity. This statement has been made many times as the opening statement when those of us who planned and carried out the IGY were called upon to speak to audiences that ranged in scientific training from the members of the Governing Board of the National Academy of Sciences to 5,000 high school students and teachers brought together by the Frontiers of Science Foundation in Oklahoma City, or to intelligent laymen interested in the times in which they live. To understand the real significance of this statement would take several chapters devoted to the sun and its complex relationship to the earth. Here, I intend to enlarge on this statement, but certainly not to treat it exhaustively. Since my own active scientific career, mostly spent at UCLA, has overlapped some of the most exciting developments in our understanding of the sun, I hope that you will allow me occasionally to become somewhat personal in this account of the sun and the earth.

I also want to emphasize that for a long time man will be exploring and studying his solar system, using the new tools and techniques that space exploration has made possible and conceivable, and that sun-earth relationships now take on a meaning that no one would have envisaged a quarter of a century ago. Since the sun is the center and controlling element in our solar system, a thorough study of these relationships will be rewarding in itself, but it will also give us information and guidance for the larger task of extending man's knowledge and activities far beyond his own planet and its immediate environment.

We could begin with the sun itself and describe some of its outstanding properties as we have learned about them from the somewhat limited viewpoint possible from the surface of the earth. We could then describe how this knowledge has been so remarkably extended during the very few years of rocket and satellite studies that have so far been carried out.

To save time, I will assume that all of you are reasonably familiar with those general properties of the sun known to astronomers and physicists for many years. I will confine the rest of my discussion to some of the new information about the sun and its terrestrial effects that has come out of the rocket and satellite programs of the past dozen years, particularly those that were related to the IGY. Even here, I will have to make a severe selection in order to avoid undue length.

I want to mention briefly the notable rocket-borne spectrographic results on the ultraviolet spectrum of the sun obtained by Richard Tousey and his collaborators at the Naval Research Laboratory. These were striking because for the first time in man's history he had a direct picture of the nature of solar radiation in regions of the spectrum not accessible to him before the use of rocket-borne instrumentation for solar research. Guesses regarding the character and intensity of the ultraviolet light needed to produce the various regions of the ionosphere and the chemical properties of the high atmosphere, were rapidly replaced by real data. With these came better knowledge of the sun, as well as of its terrestrial effects. Numbers began to replace the guesses.

Another example of this replacement of guesses by numbers is given in the following account of rocket observation of solar activity, based on a report in IGY Bulletin Number 32, February, 1960. It in turn was based on material supplied by Herbert Friedman, T. A. Chubb, and R. W. Kreplin of the Naval Research Laboratory.

As part of its contribution to the IGY and IGC-59, the NRL conducted a number of solar rocket-astronomy experiments. The most recent of these was called Sunflare II, and in this one 12 Nike-Asp rockets were launched from Point Arguello on the Pacific Missile Range between July 14 and September 1, 1959. Sunflare I, the IGY Rocket Flare Patrol Program, was conducted at Point Mugu, California, in July and August, 1957. Both Sunflare programs represented continuation of the NRL "rockoon" programs of solar observation conducted in the summer of 1956.

The objective of the Sunflare II experiment was to obtain further direct information about the nature of solar flares—in particular, about the radiations emitted by the flares as compared with those emitted by the quiet sun. Of the 12 Sunflare II launchings, eight were completely successful, reaching heights between 130 and 150 miles; in the best flights, data were telemetered continuously for nearly 8 minutes. Three of the eight rockets were launched during flares and five were launched when the sun was quiet or only mildly active.

Now a word about flares and related phenomena. A solar flare is usually seen as a sudden brightening of the sun's surface near a sunspot. A flare reaches its peak brightness—as much as ten times that of the surrounding photosphere when viewed in $H\alpha$, the red light of hydrogen— within a few minutes, during which a large flare may expand over an area of the sun's surface ranging from 100 million to 1 billion square miles. The flare's decay is slower, ranging from a half hour to several hours, depending on its size. Visible flares are flat and extend parallel to the sun's surface, remaining essentially stationary throughout their lifetimes. They are three to four times as thick as the chromosphere and their uppermost portions reach into the corona. The chromosphere, which is the lowest level of the solar atmosphere, extends to a height of about 15,000 miles above the surface. Flares grow outward at speeds only rarely exceeding about 10 miles per second.

Flares are one of a group of phenomena constituting solar activity, which directly affect the earth's upper atmosphere, frequently interfering with radio communications. The terrestrial effects associated with flares fall into two categories: (1) those occurring almost immediately following the beginning of a flare, such as radio fadeouts or sudden ionospheric disturbances (SID), and caused by solar X rays; and (2) those beginning about a day later, such as geomagnetic storms, as a result of the arrival at the earth of the high-speed stream of charged particles ejected by the flare.

The observation of the sun in visible light from ground-based stations for the purpose of forecasting SID and magnetic storms suffers from the fact that relatively minor disturbances in visible light may be followed by great terrestrial disturbances, while apparently large solar events may produce only minor effects here on earth. The direct observation of the sun's energetic radiations by rocket-astronomy techniques must therefore be a great step toward the improvement of our understanding of the physical nature of flares and of the resulting terrestrial phenomena.

Two types of ionization chambers were included in the 55-pound instrument package flown in Sunflare II. One type, fitted with a Mylar window and filled with dry nitrogen, was designed to study 44 to 60 angstrom X-ray intensities. The other, fitted with a lithium fluoride window and filled with nitric oxide, was designed to study Lyman-c radiation at 1,216 A. Two types of photon counters were also used, one with a beryllium window and one with an aluminum window. The former was designed to cover the 2 to 8 A region and the latter the 8 to 18 A region of the X-ray spectrum. Finally, two types of scintillation counters, or proportional counters, were used to study the intensity of penetrating X rays. One was sensitive to the 2 to 20 kev range and the other to the 20 to 200 kev range.

In order to be able to launch a rocket as soon as possible after a flare started, both optical and radio methods for flare detection were used. An NRL team at the Lockheed Solar Observatory in Los Angeles observed the sun's disk in the light of H_α, transmitting their observations to the launching site by private telephone. Meanwhile, radio observations of flare effects on the ionosphere were conducted from the NRL van at Point Arguello. These observations of sudden short-wave fadeout and sudden cosmic noise absorption, at a frequency of 18 megacycles per second, were reliable indicators of flares of sufficient intensity to cause ionospheric disturbances.

The indication that there existed a background of X rays in the upper atmosphere came in NRL's pre-IGY rockoon program. Here the energies appeared to be as high as 50 kev. However, because of the small flux, it was not possible to identify the radiation with the sun.

The Sunflare I measurements during the IGY period detected intense flashes of 6 to 7 kev X rays accompanying solar flares. These flashes were of sufficiently short wavelength to penetrate to the base of the D region of the ionosphere and produce the D-region ionization required to explain radio fadeouts. It was proposed that this production of X rays in the solar atmosphere could result from heating of the corona above the flare to temperatures greater than 10,000,000°C. The temperature of the quiescent corona is about 500,000°C.

The Sunflare II experiment provided further important information with the discovery of the emission of extremely energetic X rays during the most active phases of large flares. In the early phases of the three such flares observed, X rays with energy as high as 90 kev were detected. Moreover, these X rays were not emitted as brief bursts, but persisted throughout the full 6 minutes the rockets were above the absorbing region of the atmosphere (about 45 kilometers, the height to which the hardest X rays penetrated).

The best explanation of the observed spectral distribution of the X rays is to postulate the existence of a collection of thermal sources in the solar atmosphere, and the production of 80 to 90 kev X rays implies local temperatures—in the vicinity of the flare—of 100,000,000°C. When the few available measurements of solar X rays from class 2 and 3 flares are averaged, the following correlations of flux with wavelength are obtained: for wavelengths of 20 to 100 A, the flux is 2 ergs/cm^2/sec; for 8 to 20 A, the flux is 0.03 ergs/cm^2/sec; and for 2 to 8 A, it is 0.01 ergs/cm^2/sec. The total X-ray flux affecting the E region during the class 2+ flare was roughly twice the quiet sun value.

The Sunflare II results indicate that high-energy X rays are normal solar emission. It appears that even the quiet sun emits a broad spectrum of X rays extending to very high energies, although the flux is very low, and that the entire spectral range of X-ray emission is enhanced by the intense excitation accompanying a flare.

Needless to say, these observations on high-energy X rays from the sun provide the basis for much theoretical speculation as to the mechanism of their generation. I don't intend to review these here, but I do want to point out that these new observations provide another clue to the complex electromagnetic fluid phenomena that make up the physics of the sun. As such alone, apart from their important terrestrial effects, X rays from the sun represent a major step forward in observational astronomy.

An achievement in solar physics that is as notable as the one that I have just discussed in some detail is the photography of the sun's disk in the light of the Lyman-alpha line of hydrogen. From the ground or from an observatory high on a mountain top, there is no difficulty in revealing one of the sun's many faces by photographing it in the light of the visible red line of hydrogen H_α, or in the violet light of calcium. Since each of these radiations originates at a different level of the sun's atmosphere, the resulting pictures look different.

Lyman-alpha light is in the extreme ultraviolet region of the spectrum, midway between visible light and X rays. It originates in regions of the solar atmosphere 4,000 to 6,000 miles above the sun's surface. When large amounts are absorbed by the D region of the earth's ionosphere, this radiation may have a powerful effect on radio communications, causing disruption or changing the quality of transmissions on many short-wave radio wavelengths.

The photographs of the sun's disk in the light of Lyman-alpha were taken from an altitude of 123 miles by a small, specially constructed solar observatory in the nose of an Aerobee-Hi rocket launched from White Sands, New Mexico, on March 13, 1959. The experiment was designed and the launching was conducted by the Naval Research Laboratory as part of the IGY Solar Activity Program.

The importance of this achievement is so great that a description of the experiment in some detail is worthwhile. Sixty photographs were obtained within a period of about a minute, and these show great, bright, irregular clouds of hydrogen gas high in the solar atmosphere. The clouds cover about ⅓ of the disk, mostly in the sun's northern hemisphere, and have temperatures in excess of 6,000°C.

During the rocket flight on which these Lyman-alpha photographs were made, hydrogen red-line and calcium K-line photographs were also being made from the ground at a group of solar observatories in California, Michigan, New Mexico, and Washington. By comparing solar phenomena mapped at these three wavelengths, one might be able to throw new light on the process by which solar energy released inside the sun by nuclear reactions reaches the surface and escapes.

Calcium K-line photographs map the sun's atmosphere from the surface to a height of 4,000 miles, and photographs in the red line of hydrogen show the pattern from the surface to 200 miles. The photographs in Lyman-alpha show the solar weather pattern at the highest level in the solar atmosphere as yet studied over the entire surface of the sun. Thus, photographs at these three levels, when analyzed together, give a sort of three dimensional picture of the processes taking place in the sun's atmosphere.

It is apparent that the sun is strikingly stormy where viewed by the extreme ultraviolet light of hydrogen. The same bright disturbed areas are present also in the photographs taken from the earth's surface, but they are smaller and not as conspicuous. The pattern appears to become coarser at higher levels. Though in part, perhaps, ascribable to instrumental effects, this agrees with the concept of ascending columns of turbulent gas spreading, combining, and merging as they stream out of the sun. Eventually, traces of these gases reach the earth. Occasionally, during periods of great solar activity, they produce auroras and magnetic storms and disrupt radio communications.

The instrumentation for this experiment took 4 years to develop at the Naval Research Laboratory. Since Lyman-alpha is strongly absorbed by all materials, lenses could not be used and it was necessary to construct the entire camera with mirrors. These were not ordinary mirrors, however, but diffraction gratings: mirror surfaces ruled with 15,000 lines to the inch. These rulings caused the intense light from the sun to be thrown out of the camera, leaving only the monochromatic Lyman-alpha radiation to form the solar image.

Ordinary film cannot be used to photograph this image because the gelatine binder in the emulsion absorbs the extreme ultraviolet radiation before it can reach the sensitive silver halide grains. A special film containing almost no gelatine must be used. This film is very fragile and care must be taken not to touch the surface or the image will be wiped off.

Once the equipment is in flight, it is still a difficult problem to ensure the taking of good pictures because the rocket's flight path is certain to be unstable—rolling, pitching, and yawing. Precise pointing is essential, as it is to any astronomical telescope. To keep the entire instrument pointed at the sun, a complicated servo system, constructed by physicists at the University of Colorado, was employed. The spectrograph used weighed about 35 pounds and the entire payload about 250 pounds.

With these refined techniques, evidences of weather in different parts
of the solar atmosphere may be seen and attempts made to correlate
the more violent changes in solar weather with terrestrial ionospheric
weather and, possibly, with local earth weather. I have already called
attention to the fact that solar changes observed with visible light are
only indirect indications of what is happening in the more energetic,
invisible ultraviolet and X-ray emissions. Photography of the sun in
wavelengths of these powerful short-wave radiations makes possible a
tremendous advance toward understanding the relationship between
solar activity and its terrestrial consequences. Lyman-alpha pictures such
as those obtained in the March 13, 1959, rocket flight show that the
radiation is emitted from spots and patches over the solar surface.
This knowledge may help us to understand the reasons for the wide
range of variations and the rapid changes in quality of radio propagation. Eventually, such rocket astronomy techniques, leading to the discovery of phenomena forbidden to earth-bound astronomers, may permit
routine photography of the sun in the most important short-wave
radiations, providing daily solar weather reports for prediction of terrestrial responses.

And now we come to another great achievement in the study of the
sun. Recently, Dr. Herbert Friedman, whose name will be long remem-
bered as a pioneer in rocket astronomy, announced that an X-ray photograph of the sun had been taken from a rocket fired 130 miles above the
earth, high enough to be above even the thin atmosphere that absorbs
X rays.

Here too, a special technique had to be used because no lens system
could focus X rays. This time, however, use was made of the simplest
principle of photography—that of a pin-hole camera. A hole 0.005
inch wide, covered with an extremely thin aluminum film to screen
out visible light, was all that was needed. The picture shows that the
principal source of X rays reaching the earth is the low-density corona
surrounding the sun.

So we have learned to take pictures of the sun in the entire range of
frequencies, from H_α to X rays. Each tells its own story—together they
challenge our imaginations.

I can mention only briefly another interesting series of IGY experiments, again carried out by Herbert Friedman and his NRL colleagues,
during the October 12, 1958, total eclipse of the sun. This US-IGY

expedition went to the Danger Islands in the South Seas and it was the most elaborate such expedition ever mounted. Unfortunately, a beautifully prepared set of optical experiments stood idly by under the cloud-filled sky, but the rocket experiments could ignore this unfortunate situation. Such is the power of the rocket and satellite, and the weakness of the ground-based observer.

The idea behind the expedition was to design rocket experiments to measure X-ray and ultraviolet emissions from those portions of the sun which were uncovered during various stages of the eclipse. As an eclipse progresses, it is possible to measure the contributions of sunspot regions as they are covered or uncovered. During the totality, the photosphere is completely covered, so any residual radiation must come from the outermost layers of the chromosphere and from the corona. The observations by the NRL group were complemented by the observations of the group from the Central Radio Propagation Laboratory of the National Bureau of Standards, which planned to obtain data on heights, ionization densities, and critical frequencies of the different ionosphere layers.

Because these layers are ionized by solar ultraviolet and X radiation, a solar eclipse causes a marked decrease in ionospheric density. Correlation of the radiation data gathered by the rockets and the ionospheric data gathered by conventional methods should give much information about the changes of ionospheric structure with changes in solar radiation.

These experiments, correlating rocket and ionospheric observations during a solar eclipse, will undoubtedly be tried again. The one carried out during the IGY is but another example of the pioneering character of the program, demonstrated again and again in each of its many disciplines.

I have confined my remarks up to this point to problems of the sun and the earth where it is generally agreed that the phenomena are under solar control. The principal ones are the aurora, the airglow, magnetic variations, radio disturbances, and now, on a much wider scale, the radiation regions that were discovered by the first successful IGY satellite.

I want next to mention the possible meteorological effects of variable energy fluxes in auroral latitudes, and here I refer to the corpuscular radiation from the sun which is associated with the production of auroras in high latitudes. The possible meteorological effects of these variable energy fluxes have been looked at from time to time as more has been learned about their character and magnitudes. In 1956 this problem was summarized as follows by Sydney Chapman, whose vision and leadership played such a great role in the origination and carrying out of the IGY. Professor Chapman said:

> The areas and intensities and heights of the different flux components of energy associated with the aurora may appear unpromising meteorologically, even allowing for atmospheric instabilities. But I would think it regrettable if these considerations should discourage meteorologists from the search for relationships between weather and solar change. The upper atmosphere is a region very imperfectly explored and understood, and the future may still have important surprises to disclose, some of which may be significant in this connection.

I would comment here and agree with Professor Chapman and note with some personal satisfaction that the eyes of the meteorologist have turned upward and that he now thinks in terms of the sun and the earth and of heights to which he had not looked earlier. The IGY took the meteorologist to the polar regions, both north and south, and gave him a better look at events on a world-wide scale and in three dimensions. Out of this are coming new ideas and new methods for meteorology; and with this remark, I want to move into the meteorologist's domain and talk about the satellite experiment which man will some day look back at as the experiment that changed the science of meteorology—I refer here to TIROS.

Because of the historical role of TIROS in the field of weather forecasting, I am going to describe it in detail here. Dr. Neiburger, in his part of this chapter, will discuss the usefulness of the TIROS data in meteorology. I would like first to point out that a meteorological satellite was one of the most important parts of an original IGY satellite program, and because of the ease with which most people could recognize its potential significance, its inclusion did much to point out the immediate economic significance of space science.

The name TIROS is derived from the fact that the meteorological satellite, launched on April 1, 1960, from Cape Canaveral, Florida, was a television and infrared observation satellite. At the time of its launching, it was by far the heaviest yet launched by the United States and the one that has come nearest to achieving its intended orbit. The intended orbit was to have been circular at about 400 nautical miles above the earth's surface, and the actual orbit came remarkably close to this, differing in apogee and perigee by only about 28 nautical miles (33 statute miles). Its orbit extends from about 48° north to 48° south and its orbital time is a little over 99 minutes.

The satellite looks like a large hatbox, 42 inches in diameter and 19 inches high. Its top and sides are almost completely covered by banks of solar cells—some 9,200 in all. Its primary mission was to send back pictures of the cloud cover. It had two television cameras capable of taking pictures every 30 seconds for 32 exposures. The 500-line pictures were stored on magnetic tape and read off on demand from ground control stations at Fort Monmouth, New Jersey, and Kaena Point, Hawaii.

In addition to its TV cameras and associated equipment, the satellite contained beacon transmitters, attitude sensors, and telemetry circuits. Power was supplied by nickel-cadmium batteries charged by solar cells. Power output was expected to average about 19 watts.

TIROS was spin-stabilized and hence it was not looking at the earth at all times. It was possible, on the basis of tracking information, to program the cameras to take photographs only at times when the satellite was viewing the earth and when the area to be photographed was in sunlight. Pictures taken while TIROS was out of the range of the ground stations were stored on tape for later relay. Each read-out wiped the tape clean. When the satellite was within range of a station, ground command would turn on the cameras directly, and photographs taken above the station were immediately relayed to earth, bypassing the magnetic tape.

I was particularly pleased to learn that the second stage of the launching vehicle was powered by a liquid-fueled engine which was adapted and modified from earlier Vanguard and Thor-Able rocket vehicles. The third stage was a solid-propellant rocket adapted from the Vanguard and Able I rocket vehicles. These facts show the importance of the so-called "ill-fated" Vanguard system for later launchings.

Many government and industrial organizations cooperated in this program, but the one that intrigues me is the Meteorological Satellite Section of the Weather Bureau, which is responsible for analyzing and interpreting the cloud-cover data. Who would have dreamed a few years ago that the Weather Bureau would have a Meteorological Satellite Section? What is next?

You will note that I have used the word "weather" in talking about the various regions of the sun's atmosphere and in talking about the earth's high atmosphere. I like the use of the word "weather" in these connections, because it carries with it the implication that the science of meteorology, faced with the gigantic task of solving the dynamics of the earth's sensible atmosphere, will now take on a new dimension as its theoretical techniques are applied to problems of motion in the atmosphere of the sun and in the high atmosphere of the earth. Our newly won abilities to look in detail at the character of these motions provides a challenge to meteorology—one that I hope it will be able to meet.

7

Part II. Utilization of Space Vehicles for Weather Prediction and Control

MORRIS NEIBURGER

PROFESSOR AND CHAIRMAN, DEPARTMENT OF METEOROLOGY
UNIVERSITY OF CALIFORNIA, LOS ANGELES

Ph.D. (Meteorology), University of Chicago, 1945. Held various positions with U.S. Weather Bureau, 1930 to 1940; Instructor, Graduate School, U.S. Department of Agriculture, 1939 to 1940; Massachusetts Institute of Technology, 1940 to 1941; joined the staff of the Department of Meteorology, University of California, Los Angeles, in 1941 where he has served as instructor through the various ranks to professor in 1954 and chairman of the department in 1956. Member of various meteorological societies. Main interests are physics of clouds, air pollution, synoptic meteorology, upper wind dynamics and temperature forecasting, atmospheric radiation, evaporation. Received Meisinger Award in 1946.

IN FANTASY we can visualize an "ideal" weather service: a group of satellites appropriately spaced, skimming above the atmosphere, would radio back reports of cloud and precipitation distribution, and radiation measurements from which temperatures, humidity, and ozone content at various levels within the atmosphere would be deduced. Other satellites about 22,000 miles out would move around the equator at the speed the earth is turning, thus remaining fixed with respect to the earth's surface and continuously monitoring large areas. A station on the natural moon and one placed at about the same distance on the opposite side of the earth would make over-all observations of the total radiation emitted by the earth.

The data from these satellites, together with those from automatic observing stations distributed over the earth's surface, would be fed automatically into data-storage and processing machines at a few strategically located centers. The prediction of the world-wide atmospheric circulation patterns would be carried out at one or two computing centers. At these centers machines would automatically take data from the storage centers as needed and would compute forecasts for weeks and months in advance, perhaps even for years. The computing machines would continuously check the previous forecasts against the observations, and whenever the differences became significant they would recompute and correct the forecasts for the future on the basis of the new observations.

At regional computing centers machines would automatically take the world-wide forecasts and from them compute the specific forecasts of temperatures, precipitation, winds, etc., for their regions. At flight forecasting centers, machines would evaluate optimum flight paths, estimated flight times, launching or take-off conditions, and winds and weather conditions in flight. These would be computed automatically for a specified program of scheduled flights and computed on demand for unscheduled flights.

The output from the forecasting machines would go into a communications network from which there would be continuous printout of information as desired at airline (or spaceline) offices, industrial establishments where operations or planning required weather information, newspaper offices, and television stations. And the general public could get the up-to-the-minute forecast by dialing the telephone or television.

Forecast utilization centers would carry out computations on such matters as the effect of the forecast weather on transportation, automobile traffic, and the requirements or demand for water, food, clothing and other commodities, so that appropriate measures to meet these needs could be taken. For instance, the forecast machines might be programmed to adjust the traffic signals on the basis of the effect of the weather forecast on automobile traffic, and similarly, they might compute the length of time and how often that orchard heaters would be required for fruit protection, and be programmed to turn on and off these heaters.

At weather modification centers, machines would take the forecasts and on the basis of predetermined programs and criteria compute the desirability and feasibility of altering unfavorable weather that is forecast and, if desirable and feasible, the optimum way of modifying it. The computations would include all the consequences of the proposed modification, so that the decision to modify for the benefit of some would not adversely affect others. Some phenomena, such as severe windstorms and tornadoes, flood-producing thaws or rains, or crop-devastating droughts would always be modified if possible; but even for these the question of whether the attempt to minimize them might lead to phenomena causing greater losses would have to be considered. All these considerations would automatically be made by the computers, so that the answer would point up clearly the optimum decision.

How close are we to such an ideal system? Clearly many problems other than that of the observing platforms supplied by satellites and other space vehicles are involved. We shall not concern ourselves with these and shall look only at the question of how much these vehicles will contribute.

The utilization of space vehicles for weather prediction and control seems so obviously promising that many rash forecasts of great improvements have been made from the time such vehicles were first envisioned. I say rash, for even persons of adequate competence to appraise the problem often have not considered it in sufficient detail, but rather have engaged in vague generalities. The scientists who have studied the matter with care have been much more cautious. They have suggested that there will be slow improvement in weather forecasting during many years of learning how to utilize the data from satellites as well as how to solve the prior problems of instrumentation of the vehicles for the accurate measurements required and of launching them in orbits of the precise nature needed for meteorologically optimum data.

It is easy to see that a vehicle orbiting frequently around the earth will pass over large uninhabited areas—oceanic, desert, or polar—where weather observations are absent, and it thus could detect storms which would otherwise not appear on the meteorologists' charts until they came close enough to an observing station to affect it. Successive traverses over such a storm—perhaps a typhoon in the Pacific Ocean— would enable the forecaster to tell which way and how fast it is moving, and thus issue warnings to ships, islands, or coastal areas earlier than he would be able to if he were dependent solely on earth-bound vehicles. Even the use of reconnaissance aircraft has not always located typhoons and hurricanes at early enough stages, and their use has involved risk of the lives of the crew—a factor that would not be involved if the information were telemetered from unmanned vehicles.

Already a start in this direction has been made, with the launching of the TIROS I satellite in April, 1960. This satellite was equipped with two television cameras to record the cloud distribution for transmission back to earth by radio. Approximately 23,000 photographs of cloud systems from various parts of the earth were received before transmissions were stopped in the latter part of June because some of the components were misbehaving. These cloud pictures can be interpreted in terms of the kinds of weather systems which produced them.

There are other immediately obvious benefits to weather forecasting which would ensue from the use of adequately instrumented satellites, which I shall discuss later. First, however, I should like to explain some limitations to their utility, and some reasons why we cannot look forward to immediate and complete solution of the weather-forecasting problem as soon as we start surveying the earth from satellites regularly.

It has long, and with justice, been the complaint of the weather forecaster that he does not have enough data. He does not have all the data required for location of the weather systems which will move to his area in the next few hours or days, nor enough data at upper levels to determine accurately the dynamics of the atmospheric flow patterns which control the way the systems will move and change. The requirements for better weather information have led in recent years to a large increase both laterally and vertically in the observational network, but even today only about one-fifth of the atmospheric mass is adequately probed. While it must be acknowledged that there has been improvement in accuracy and detail in the forecasts, this improvement has been small compared to the tremendous increase in number of observations. We have reached the state where only a fraction of the observations can be processed and used even at the largest meteorological analysis centers. The numerical weather prediction unit at the National Weather Center of the U.S. Weather Bureau in Suitland, Maryland, which uses an IBM 7090 electronic digital computer for analysis of the data and preparation of prognostic charts of atmospheric flow patterns over the Northern Hemisphere (like its neighboring weather map analysis center, which analyzes manually the weather situation in three dimensions over the hemisphere), must use a selection of the data available because the problems of transmission, processing, and analysis limit the number of data which can be handled in time for use in the forecasts.

Thus only in so far as the data from satellites will be of better quality or greater value, and therefore be worth displacing existing observations as input in the forecasting process, can they be immediately beneficial. But, in fact, the kinds of observations that can be made from outside the atmosphere are largely incompatible with the existing forecast systems, in which the primary parameters are the pressure, temperature, humidity, and velocity of the air at discrete points within the atmosphere. Already studies are under way concerning means by which the extraterrestrial observations can be incorporated into present forecasting procedures. However, by and large we must anticipate that their utilization will involve the development of completely new procedures and a complete revision of the use of the more conventional observations.

The limitation arising from the types of data which can be obtained from outside the atmosphere becomes clear when we consider that our knowledge of the atmospheres of Mars and Venus, for instance, is very meager. We look forward eagerly to the time when space vehicles will visit these planets and send back more information about their nature. A look at the earth's atmosphere from the surface of these planets would be of little or no value. From distances of the order of that of our moon and closer, the situation begins to be more favorable, giving the opportunity to view in some detail up to one-half of the earth at a time, and the entire earth in a period of the order of one day. But since the information collected even by a satellite skimming the edge of the atmosphere is always a form of electromagnetic radiation (such as the solar radiation reflected or the long-wave temperature radiation emitted by the earth's surface and the atmosphere, or radar signals emitted by the vehicle and reflected by the ground or clouds), it is restricted to the conclusions which can be drawn from such probing from a distance. As we will see, for the most part this information at best can supplement, but not replace, the observations and soundings made at the earth's surface and within the atmosphere.

The data-handling difficulties mentioned will of course be overcome by development of larger and faster machines; and if the kinds of data to be obtained from space vehicles have their limitations, they nevertheless will constitute means of supplying some information concerning the large fraction of the atmosphere not presently adequately probed.

Having made it clear that the introduction of data from space vehicles will not produce an overnight change in the accuracy of weather forecasts, I can now turn again to the positive side and discuss some of the contributions we can look forward to as the appropriate data become available and we learn how to use them.

First and most important is the radiant energy from the sun and its variations. The atmosphere is the thermodynamic fluid of an engine which ultimately derives practically all its energy from the sun. While the response is by no means as immediate and direct as the response of an automobile engine to the change in fuel supply when the accelerator throttle is varied, it is clear that variations in solar energy must have some influence on the weather. How much influence—is the effect amplified or is it damped—is a question which has been the subject of much controversy for decades, in part because of the uncertainty in the amount and character of variation of solar energy. Having a station outside the atmosphere to measure continuously the radiation from the sun in all wavelengths, and also corpuscular emissions, will settle the question of the magnitude of these variations. At present the best estimates are that the total variation of electromagnetic wave radiation is less than three-tenths of one per cent, but it has been suggested that corpuscular emissions may augment the energy received from the sun by as much as 10 per cent at times, near the latitudes of maximum auroral frequency. While these corpuscular emissions are absorbed in the very high atmosphere (always higher than 40 miles), it is conceivable that they might subsequently make their influence felt at lower levels. Tracing and testing possible connections would be greatly facilitated by continuous records of the intensity of the emissions leaving the sun and reaching the outer limit of our atmosphere. If it becomes established that variations of adequate magnitude to have effects in the lower atmosphere do occur; if the connections between these variations and the effects become well understood; and if there is sufficient time lag between the occurrence of the variation in solar energy and the weather effect caused by it, then the observation of solar energy from space vehicles will give an essential key to weather changes not presently possible to forecast.

Another cause of variation of the energy input into the atmospheric engine is changes in the reflection of the incoming solar energy by the earth's surface (especially that covered by snow), by clouds, and by dust and haze in the air. For instance, the reflectivity of clouds, although depending on their thickness and other factors, may average about 70 per cent. If from one day to the next the fraction of the earth's surface covered by clouds changed by 5 per cent, the change in energy available to drive the·atmospheric circulation would be 3½ per cent, ten times the presently estimated maximum variation in the solar constant; and this variation would be entirely in that portion of the energy which penetrates to the lower layers of the atmosphere. Variations due to the change in amount of surface covered by snow, and in the reflectivity of the snow due to aging, may be of similar magnitude. Measurements from satellites of the radiation reflected back from the earth to space would thus give another essential piece of information concerning the energetics of the atmosphere which is not presently available.

A third aspect of the energy budget of the atmosphere is the outgoing infrared radiation from the earth and the atmosphere. The earth's surface radiates almost as a black body, as do the clouds where they are present. Part of the energy radiated by them is absorbed by the water vapor, carbon dioxide, and ozone in the atmosphere, and these gases in turn radiate at their own temperatures in the wavelengths in which they absorb. The net radiation leaving the outer edge of the atmosphere depends not only on the temperature distribution at the earth's surface and at the cloud tops where there are clouds, but also on the distribution of temperature, water vapor, carbon dioxide, and ozone throughout the atmosphere. To estimate the energy loss to space would require extensive detailed measurements not presently available, but measurements of this loss would be relatively simple from a satellite.

In addition to measurements of these quantities for the earth as a whole, which could be made from satellites in the range 20,000 to 200,000 miles from the earth, or from the moon, measurements would be made for small areas from satellites in the range 300 to 3,000 miles. Those places where incoming exceeds outgoing energy, and where energy is thus being stored, would be differentiated from those where the reverse is true. Daily determination of this distribution of energy sources and sinks would enter into the computation of the atmospheric circulation, which must in the end result in transfer of energy from the places where it is stored to the places where it is lost.

Preliminary measurements of the radiation budget of the earth and atmosphere have already been made by the Explorer VII satellite (1959 Iota), which was launched in October, 1959, as part of the IGY program. This satellite was equipped with six sensing elements for radiation; three measured total energy at all wavelengths from various directions, one measured long-wave (infrared) radiation, and two measured short-wave (solar) radiation.

It is the measurement of the distribution of reflected energy which gives the possibility of determining the distribution of clouds, and thus of cloud-producing storms. In general, the reflectivity of clouds is much greater than that of the earth's surface, so that it is possible to distinguish them by the contrast in reflectivity. An example of this is shown in Figure 35, which presents a photograph taken from an Atlas missile on August 24, 1959, at an elevation of 636 miles. On this picture the white areas are clouds, the black are ocean surface. The broad cloudy band across the right center of the picture is at a front between two air masses, with waves on it. Figure 36 shows the result of piecing together a number of pictures taken on the Atlas flight. As the missile turned, the camera viewed in various directions. The frontal clouds shown in the previous figure are at the top; this band of clouds extended eastward from South Carolina across the North Atlantic Ocean. On the right of the picture are the tropical convergence zone clouds (repeated

FIGURE 35. Photograph taken from Atlas missile at an elevation of 636 miles.

in the lower left), which extend eastward from the northern coast of South America.

In the first attempt to obtain cloud data from a satellite (Vanguard II), photoelectric cells were used to sense the reflected radiation. For this type of receptor it is necessary to have very precise motions of the satellite to be able to put the data together into pictures such as those shown. Unfortunately, Vanguard II had an unsuitable orbit and a bad wobble, so that the data received from it could not be put together in this fashion. The next meteorological satellite, TIROS I, used television cameras directed along the axis of spin of the satellite, and thus gave cloud pictures whenever the axis pointed toward the sunlit portion of the earth, independently of the rate of spin or the wobble of the axis. The wide-angle camera, with a field of 104 degrees, when looking directly downward gave useful information over a large area, a square 750 miles on a side; the narrow-angle lens, with a field of 12.7 degrees, gave details of the cloud structure over a much smaller area, 65 miles square, when it pointed vertically downward. Actually the cameras were directed ver-

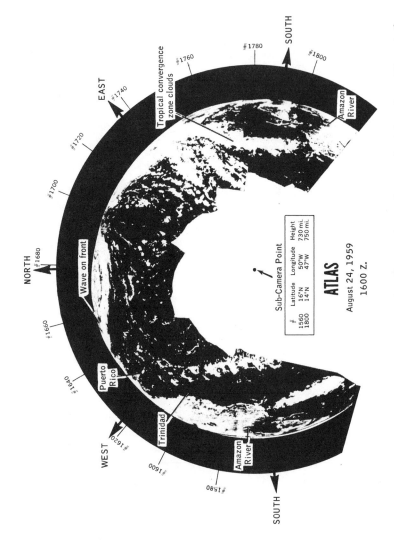

EAST

NORTH

WEST

SOUTH

SOUTH

#1740

#1720

#1700

#1680

#1660

#1640

#1620

#1600

#1580

#1760

#1780

#1800

Tropical convergence zone clouds

Wave on front

Puerto Rico

Trinidad

Amazon River

Amazon River

Sub-Camera Point			
#	Latitude	Longitude	Height
1560	16°N	50°W	730 mi.
1800	14°N	47°W	750 mi.

ATLAS

August 24, 1959
1600 Z.

FIGURE 36. Composite, made by U.S. Weather Bureau from photographs taken from Atlas missile.

162

tically downward only rarely, and most of the time they gave distorted pictures of the cloud pattern over much larger areas.

Figure 37 gives an impression of the area covered by the wide-angle lens; the Red Sea is in the lower right of the picture, the Sinai peninsula in the upper center, and the eastern Mediterranean Sea in the upper left.

In Figures 38 and 39 the contrast between the wide- and narrow-angle cameras is shown. Figure 38, taken with the wide-angle camera, shows the banded clouds of part of a cyclonic storm over the central Pacific Ocean; the storm is several thousand miles in diameter. Figure 39, taken with the narrow-angle camera 1 minute later, gives details of the cloud structure in the center of Figure 38.

FIGURE 37. Photograph transmitted from meteorological satellite, Tiros I. showing area covered by wide-angle camera.

Figure 40 shows the correspondence between cloud systems photographed by TIROS I and an analyzed weather map. The storm center, marked "low" on the map, is in the circular cloud mass in the upper part of the picture. The cloud extending southward from center is in the warm, moist air mass between the two fronts shown by heavy lines in the diagram. It is completely cloudless to the west of the cold front, which is the dark line with black triangles along it. The photographs in

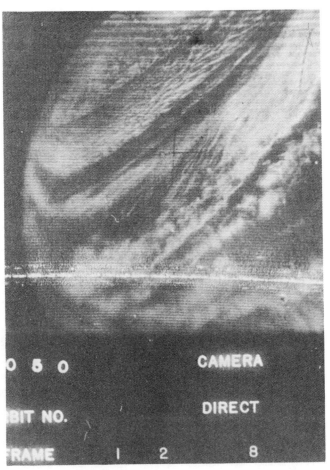

FIGURE 38. Banded clouds of extra-tropical cyclonic storm over Pacific Ocean, photographed with the wide-angle camera from Tiros I.

Figure 40 were taken by TIROS I at 3:30 P.M. on April 1, 1960, during its fifth orbit on the day it was launched.

Over snow-covered portions of the earth the reflected sunlight will not differentiate between clouds and snow-covered ground. However, the infrared radiation emitted may possibly enable determination of the cloud distribution in these areas as well as in places where it is night. Ultrasensitive television cameras may also give the possibility of

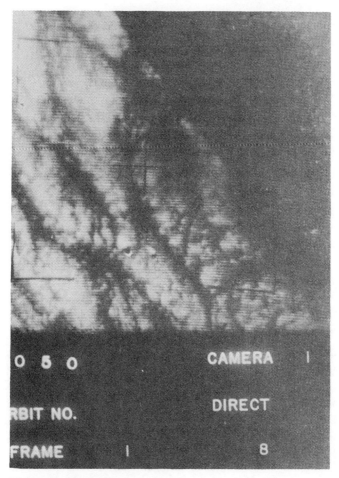

FIGURE 39. Photograph with narrow-angle lens from Tiros I, showing details of cloud structure near center of Figure 38.

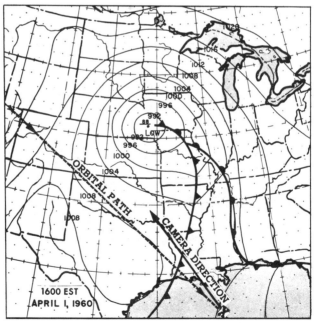

FIGURE 40. Composite cloud photograph from Tiros I, and corresponding weather map.

obtaining cloud distribution at night by means of reflected moonlight or even starlight.

In addition to the distribution of clouds over the earth, observations from satellites may be able to give the distribution of precipitation by means of radar signals, which would be reflected from the raindrops or snowflakes back to the satellites. With radar using sufficiently short waves, the distribution of clouds could be obtained; this would be particularly useful over the night-time portion of the earth where ordinary television would not be effective. Three-centimeter radar is suitable for measurement of precipitation areas, and shorter than 1-centimeter waves would be reflected from cloud particles as well.

By measuring the infrared radiation in various wavelengths from a satellite there is the possibility, at least in principle, of determining the temperature of the earth's surface and of various layers of the atmosphere, and also of determining the amount of water vapor, ozone, and carbon dioxide. To what extent this can be done in practice remains to be investigated, and it may take several years of intensive research to develop the instruments and methods of analysis required for these purposes.

Information about pressure and wind, most crucial in present methods of weather forecasting, is the hardest to come by from vehicles outside the atmosphere. The direction and character of plumes of pollution from large industrial complexes or from cities may be visible and give information regarding winds at low levels, but of course such populated regions would already have surface weather observations. Over oceans and unpopulated land areas one can get some information from the arrangement and tilt of clouds, which may be interpreted in terms of the wind direction at the levels at which they occur. The large-scale cloud distributions can similarly be interpreted in terms of the circulation patterns which cause them, and in particular they give information about the vertical motion of the air. From these estimates of wind distribution, crude estimates of the distribution of pressure can be made.

No doubt with time new techniques, presently completely unknown, will be developed to get from satellites more complete and more precise measurements of the various quantities which are needed for the weather forecast. For the present it appears that the data obtained from them will for the most part supplement the observations made from the ground, filling in gaps in places where observations currently are not available, and providing certain information about the over-all energy budget of the atmosphere.

The next step in the TIROS project, TIROS II, will be a satellite carrying not only television cameras, as in TIROS I, but also instruments for measuring infrared radiation. The subsequent meteorological satellites, to be developed under another program, Project NIMBUS, will be launched on polar orbits, and be equipped with more advanced equipment, including television, scanning and nonscanning radiation detectors, spectrometers, and radar.

I have mentioned that studies are already being made to see how the data from satellites can be incorporated into the present weather forecasting procedures. These procedures are based on the assumption that future weather developments depend on the past and present behavior of the atmosphere. In the numerical methods using machines, the present motion of the air, as represented by the pressure distribution at various levels, is inserted into certain mathematical equations by which the future patterns of wind and pressure are computed. In the premachine human forecasting techniques, the movement and development of pressure systems are estimated on the basis of past experience with similar systems and the crude qualitative application of the theory on which the mathematical equations are based. In both cases the atmosphere is treated as a self-contained, determinate system, in which the only disturbance from outside is the interaction with the earth's surface.

At present the weather forecasts are a composite of the results of the numerical methods using machines and the subjective procedures. The machine computations result in prognostic charts of the large-scale distribution of pressures and winds which are used by the forecaster as a basis for his detailed prognosis. The forecasts of temperature changes, clouds, and precipitation come from his interpretation of the forecast pressure and wind patterns by application of physical reasoning and previous experience.

The first step in incorporating satellite data will involve the interpretation of cloud data subjectively in terms of pressure and wind distributions which then can be used in the conventional manner. I have already given an example of the way the cloud system associated with a typhoon would enable the forecaster to detect the storm's presence and predict its movement. From the extent and nature of the clouds he can also estimate the stage in the storm's development and the strength and direction of the winds in various parts of it. Similarly, the cloud patterns associated with fronts and higher-latitude cyclonic storms, and with air masses of various sorts, would be analyzed by the forecaster to fill in the areas on his weather maps where data are otherwise unavailable, but from which the weather systems might move to his forecast area or influence the behavior of those which do. In the case of the numerical method, any information at all, even guesses, over the areas where there are presently no observations, will lead to considerable improvement in the forecasts. The prognostic equations are quite sensitive, and the forecast at any one point is affected to some extent by the pressure distribution everywhere else, particularly for longer time intervals.

At first this interpretation might be completely subjective, but we may expect that methods will quickly be developed to carry out the process automatically by machine. A crude method, for instance, might be to have in the memory of the machine a library of cloud systems and associated pressure distributions which have been previously observed, to have the machine search out the one which resembles most closely the cloud system observed by the satellite, and then to fit the corresponding pressure distribution as well as possible into the distribution determined for other areas by actual observations of pressure. One can conceive of more refined methods, in which the machine would use the dynamical equations to seek out the horizontal wind field which would produce the vertical velocities that must be present to produce the observed clouds.

The temperature data obtained from radiation measurements would give the subjective forecaster information about air mass properties which he could apply in forecasting temperature changes, and also in judging the probability of showers and other precipitation forms. When the machine-forecasting methods become more sophisticated, temperature will play a more important role, and the availability of temperature data from areas where it would be otherwise unknown will be as significant to the improved accuracy of the forecasts as the pressure data.

More important than their immediate contribution to the accuracy of weather forecasts is the value of the data from satellites in the research on weather processes, research which will ultimately make possible the realization of some of the hopes expressed in my description of the ideal weather service. Already the cloud pictures which have been obtained from TIROS I are giving us new insights into the structure of storms and other weather systems. When we get data on the variations of the energy input into the atmospheric engine, we can know to what extent the present forecasting methods, in which these variations are neglected, are valid. In addition to questions of changes in the over-all radiation entering the atmosphere, there are the questions of local energy surpluses and deficiencies to which I referred earlier. Once these data become available it will be possible to trace their effects and to develop ways of using them in forecasting. The introduction of radiation data from satellites may be expected to revolutionize our understanding of atmospheric processes by permitting the inclusion of the energy terms in the equations of dynamic meteorology.

In addition to giving information about the incoming radiation from the sun, rockets and satellites will provide the possibility of controlling it, at least to some extent. The largest known terrestrial influence on incoming radiation heretofore was the tremendous eruption of the volcano Krakatoa, in 1883, in which fine dust particles thrown into the stratosphere were carried around the earth for several years before falling out. This stratospheric layer of dust scattered back enough solar radiation to lower the average temperature of the earth about 1 degree. Similar effects could doubtless be achieved by sending up rockets or satellites loaded with fine powder and releasing it by a small explosion which would spread it into a cloud of dust. If it were released outside the atmosphere, the position and movement of the cloud would be determined at first by the location and movement of the carrier, and thus local effects could be produced. If the particles were released in the stratosphere, the winds would quickly spread them over the earth and global effects would be attained. By decreasing the incident radiation locally or globally, as needed, it is conceivable that the effects on the weather due to changes in the radiation coming from the sun could be counteracted, or that the atmospheric circulations which would occur normally could be altered.

The duration of the decrease in radiation, as well as its amount, could be controlled by the size of particles used: large particles would fall out rapidly giving an effect of short duration; very small particles could take a season or a year to fall out, during which their effect on the radiation would continue.

Of course the preceding suggestion is highly speculative. Our knowledge of the possibilities of weather control is very rudimentary. Except for direct local modification, such as by orchard heaters and wind generators for the prevention of frost and the use of cloud-seeding, which under favorable circumstances can produce some increase in the amount of precipitation, there is no known way to affect the weather and climate of an area. Attempts at prevention of hail and lightning and at influencing the formation or movement of tropical storms have so far been unsuccessful. Proposals for modification of larger scale aspects of the weather have taken the form of vague suggestions, such as the one given above, which would require much further analysis before their feasibility could be evaluated.

The primary contribution of space vehicles to weather modification and control, as well as to weather forecasting, may be the increase in our knowledge of atmospheric processes, without which no intelligent control measures can be taken. The global measurements of incoming and outgoing radiation will enable the determination of the effects of variations of these quantities on the atmospheric circulation and the weather. The question of whether we are inadvertently modifying the climate of the earth by the increase of the carbon dioxide content of the atmosphere from the burning of fossil fuels may also be answered by these measurements. The more detailed observations from satellites nearer the earth will enable a better understanding of the way storms form and develop, which may lead to discovery of means by which they may be stopped at an early stage or by which their paths may be altered.

Measurements of meteor dust from outside the atmosphere will settle the question of whether these dust particles serve as freezing nuclei which cause clouds to precipitate. If this is found to be so, the periods of deficiency of such nuclei of extraterrestrial origin would be the times when artificial seeding would have its greatest effect in stimulating precipitation.

If it is concluded, as present evidence suggests, that the carbon dioxide content of the air is increasing, and with it the average temperature of the earth, periodic creation of a layer of dust at high levels could be used to offset the temperature effect. The evaluation of the magnitude of the heating, together with studies of the rate at which the polar ice is receding, would determine whether the melting of the polar ice is resulting in a threat of inundation of coastal areas by raising the sea level. If it is, countermeasures such as reduction of the rate of burning fossil fuel, or introduction of stratospheric dust layers to reduce incoming solar energy, would be urgent.

The quantities of mass and energy which are involved in weather processes are tremendous when compared with most of the sources which man has at his disposal for possible modification of them; only the H bomb involves energies of the same order of magnitude. Modification of the courses of fully developed storms, if otherwise feasible, would require use of an energy source man is reluctant to use. It would be preferable from all standpoints, including the economic, to achieve the desired changes with small expenditure of energy.

The hope would be that storms could be altered in an early stage of their development, when their future hangs in balance and a small impulse on one side would tip the scale away from an undesired development. That such unstable conditions exist at times is highly probable: Many temperate zone storms begin as small waves on the boundaries between air masses, and tropical storms similarly frequently arise from small waves in the easterly trade winds. We presently have no way of recognizing which small waves are unstable and which are not, and even more we do not know what can be done to prevent the ones which are unstable from growing or to control their growth. For one thing, with the present loose network of observations over much of the earth, it is infrequent that a wave can be detected before it has begun intensifying into a storm. If satellite reconnaissance makes possible more or less continuous watch for such atmospheric waves, information can be accumulated which may enable us to determine the conditions under which the waves are unstable and the way in which the release of the instability can be prevented or controlled.

One could speculate ad infinitum on the kinds of information related to weather control possibilities which will come from satellite observations. It is clear that we are far from the stage where pressing a button at the ground will activate a ray gun on a satellite to dissipate a storm in one place or to make it rain in another. It is equally clear that there will be many discoveries in meteorology, presently undreamed of, which will come with the increasing use of satellites as platforms from which to observe atmospheric processes.

The promised contribution of space vehicles to meteorological knowledge is being pursued actively by the NASA, the Weather Bureau, and the various branches of the Department of Defense. Much of the material I have cited comes from publications and releases by these agencies and their contractors. Between the activities of these agencies and the research on the subject at universities, we can hope that soon man will have obtained a greater measure of understanding and control of the weather than is suggested by the freely translated lines of a German poet:

> It rains whene'er it wants to rain,
> In big and little drops;
> And when it wants to stop again
> It goes ahead and stops.

8

Atomic Energy and Space

WILLARD F. LIBBY

PROFESSOR OF CHEMISTRY

UNIVERSITY OF CALIFORNIA, LOS ANGELES

Ph.D., University of California, Berkeley, 1933. Instructor to Associate Professor, Department of Chemistry, University of California, Berkeley, 1933 to 1943; Guggenheim Memorial Foundation Fellow 1941, 1951, and 1959 to 1962; worked in Manhattan District project at Columbia University, 1941 to 1945; Professor of Chemistry, Institute for Nuclear Studies (now Enrico Fermi Institute for Nuclear Studies), University of Chicago, 1945 to 1954; appointed by President Eisenhower as a member of the U.S. Atomic Energy Commission, 1954 to 1959; Research Associate of the Carnegie Institution of Washington Geophysical Laboratory, 1954 to 1959; Professor of Chemistry, University of California, Los Angeles, June, 1959, to present. Has done a wide range of scientific advisory and technical consultant work with industrial firms participating in the Institute for Nuclear Studies, as well as with defense agencies, scientific organizations, and universities. Known for his work on natural carbon 14 (radiocarbon) and its application to dating of ancient archaeological artifacts, and natural tritium. Received the Research Corporation Award for 1951 for the radiocarbon dating technique, and the Nobel Prize in Chemistry for 1960. Member of American Chemical Society, and other professional societies. Member of National Academy of Sciences. Honors: D.Sc., Wesleyan University, 1955; Syracuse University, 1957; Trinity College of University of Dublin, 1957; Carnegie Institute of Technology, 1959.

SPACE TRAVEL and exploration mean three things: One must escape from the earth's gravitational field, one must be able to return to the earth, and one must be able to send messages back and forth. All three of these requirements for space travel and exploration have their energy aspects, and it is in energy that the atom excels.

Nuclear energy has the characteristic of being essentially weightless. The amount of material necessary to release the energy equivalent to thousands of tons of chemical fuel can be held in your hands. We are all familiar with this magnitude in that we have all read about it, but the comprehension of it has probably not been achieved by anyone. I mean there are some things beyond human experience and this is one of them—to hold a piece of uranium 235 in your hands, which if increased by a small amount or changed in shape by a certain amount, will unlock quantities of energy equivalent to thousands of tons of coal or oil being burned. This is a fact, but it is a new fact in human experience, and one therefore that is not known to people, and one that our ordinary senses do not prepare us for. There is nothing in the world in the way of feeling or common sense to tell you that a piece of uranium-235 metal differs from a piece of ordinary uranium metal. There is nothing about it to show you its awful power and potentiality. But it is there, and this is a fact we must learn to live with and a fact that we should try to use whenever energy in small packages or in small masses is required. And space travel does indeed require energy in concentrated form, so there is little doubt that space will be conquered by man only in nuclear powered vehicles. Atomic energy and space are wedded perfectly in the future because of the ultimate necessity of using atomic energy for our space vehicles.

It is therefore appropriate that we look at these problems in a little more detail, and talk of the immediacy of atomic energy in the space program. How important is it that we should move in this direction, and how likely is it to pay off in the next few years? There is little doubt that eventually the nuclear engine will be the answer for propulsion, for one cannot imagine payloads of a few tons or more, or missions such as moon landings and take-offs, interplanetary explorations or stationary satellites, without the use of at least a nuclear upper-stage rocket— perhaps with chemical boost stages. And if it should be possible to re-place the entire system by nuclear rockets, one can see an increase of payload relative to gross weight by factors of five to ten. As Dr. Schreiber, of the Los Alamos Scientific Laboratory, said recently in remarks pre-sented before the Joint Committee on Atomic Energy in the hearings on "Frontiers of Atomic Energy Research,"

> Very difficult missions such as rapid or extended space maneuver-ing, such as might be required to avoid high intensity radiation zones or to accomplish rapid interplanetary travel, probably can be accomplished only by means of some form of nuclear propulsion. The present levels of nuclear performance are barely adequate for such missions and the case for chemical propulsion is essentially hopeless.

Now, of course, Dr. Schreiber, being the head of the nuclear rocket propulsion project at the Los Alamos Scientific Laboratory, would be expected to be of this opinion. But there is a general feeling among all of the experts that something like this last statement is true. There is a point in the ratio of payload to gross weight where nuclear technology must enter, and this is because nuclear fuel has about 10 million times as much energy for a given weight as do chemical fuels.

Project Rover

The Los Alamos Scientific Laboratory has been engaged for the past 4 or 5 years in the development of a nuclear rocket engine. This is called Project Rover. It is a straightforward, open-cycle, heat-exchanger appli-cation where energy from a nuclear reactor is transferred to the propel-lant gas—hydrogen—which is heated and then sent out the nozzle of the rocket to produce the thrust. As we said earlier, nuclear power has the unique characteristic of low mass, if shielding for protection against radiation from the chain reaction is not required. If the firing site for a nuclear rocket is so well shielded and so remotely located that these radiations are not harmful, then in a very light reactor thousands of megawatts of power can be released over a period of several minutes. It is necessary, of course, to be sure that the fission products which are formed in the course of the nuclear reaction do not escape during this launching period.

The basic problem in all rocketry as presently developed is that, even for travel through the lower part of the atmosphere, the fuel and its oxidizer must both be contained in the vehicle, so the vehicle must initially carry the maximum mass of chemicals. With nuclear rockets the fuel requirement is removed almost completely since the ratio of 10 million between the energy derivable from a given weight of nuclear fuel and that from chemical fuel is dominant. A nuclear reactor the size of a trunk can easily produce all the energy needed for the heaviest flight. The problem is that this isn't enough, for in order to propel a rocket something must be thrown away. There is a fundamental law of nature that an object can be accelerated in a vacuum only in this way. Now it is true that the harder the throw is, the more effect for a given mass in the push produced. That is, the higher the speed with which the matter to be thrown away is ejected, the greater the thrust on the rocket. More quantitatively, what is involved is an equation which says that the thrust is proportional to the speed multiplied by the number of pounds per second which are ejected. Now mass is just what nuclear reactors, and nuclear engines in general, pride themselves in not having.

So the problem is this: The nuclear reactor must be supplemented with a source of material to be thrown away; and what can the nuclear energy eject which is more efficient in producing push on the rocket than the combustion products of an ordinary chemical fire such as burns with liquid oxygen and kerosene? This is the present question of the utilization of atomic energy for space propulsion.

Various solutions are being considered at the present time. One is the Rover Project, which is essentially the use of a slow throw of a large tonnage of inert material, taking advantage of the fact that because the material itself need not be the product of a fire—that is, it need not be the result of a chemical combustion process—it can be better chosen, so as to have more thrust for a given weight ejected. Now it is clear that since the speed of ejection is vital, the reactor must be run as hot as possible so that the molecules which are speeded up by the heat of the reactor are moving as fast as possible when they go out the nozzle. And the reactor block, therefore, has to be made of materials of the highest melting point, of which graphite is, of course, an obvious choice.

The nuclear fuel is allowed to generate heat, which is then removed by large volumes of hydrogen gas pumped through the graphite reactor block, being heated in passing through the block and leaving the rocket at high temperatures. These temperatures, in fact, approach the reactor block temperatures although, of course, they never achieve it.

Let us consider the amount of hydrogen required. If we take as an example a 1 million pound thrust rocket and start out with a temperature of the graphite reactor at something like 6000°F, some 2 or 3 tons of hydrogen per second is required to be thrown away. Now this obviously is a very major aspect of the problem of the Rover machine. In other words, is there a real advantage in using nuclear power when it is necessary to give up its characteristic of light mass in order to have something along to be ejected? The answer apparently is yes, though the advantage is not so completely obvious. The fact that hydrogen can be used rather than carbon dioxide and water, the more normal combustion products, is dominant. In fact, this is the principal advantage of the Rover engine over the chemical engine as it is developed at the present time. The point here is that at any given temperature, molecules move on the average with a certain energy, and the energy is proportional to the mass multiplied by the square of the speed. Therefore, at any given temperature, namely the temperature of the graphite reactor block (which, incidentally, is about the maximum temperature which can be attained in ordinary combustion processes, for the reason that the materials requirement is dominant), molecules will be moving with an energy corresponding to this temperature; but for two molecules of different mass, the one which is lighter will be moving more rapidly to make up for this fact, so hydrogen at the same temperature as carbon dioxide gives about 4.6 times as much thrust to the rocket. This is no minor advantage, and it appears that the nuclear rocket is worth building for this reason, in spite of the fact that although the energy supply is essentially inexhaustible, the supply of matter to be thrown away is not.

There is another problem, however, in connection with the Rover engine—one which we referred to briefly above. This is that during the launching great showers of gamma rays and neutrons are released and radioactivity is built up in the reactor block. The showers of neutrons and gamma rays, which are the materials released during the course of the reaction, are all over with, of course, as soon as the reaction stops some minutes after the switch is thrown, but the fission products remain radioactive (as fall-out material does) for matters of months and years, and therefore it is necessary to be sure that this material does not cause damage. The radioactivity is formed at high temperatures and is, therefore, somewhat difficult to contain because of the need to push the reactor to its ultimate temperature in order to attain the maximum thrust. The maximum thrust can only be gotten from the hottest gases. The hottest gases are obtainable only from the hottest reactor blocks, which of course contain the fuel, uranium 235. This requirement that the radiation from the fission products generated during launching be guarded against means that a logistic problem exists for nuclear rockets which does not exist, at least in such an aggravated form, for chemical rockets. It is essential to shield the launching pad using considerable masses of concrete or other materials, and it is necessary to be sure that the nuclear rocket will not spread its radioactivity over the countryside.

These are difficult problems—problems which the chemical engines do not have—but problems that do seem likely to be solvable. There are definite indications that progress is being made in their solution.

The Rover engine has another problem which is fairly sizable and may cause it to be delayed. This is the matter of the extreme chemical activity of the hot hydrogen gas that is issuing from the center and latter parts of the reactor. This activity causes the hydrogen to corrode the graphite passages and thus potentially to change the nuclear control characteristics of the reactor—a possibility which must be guarded against very carefully because the programmed removal of the inhibiting control rods, which decide how much energy is to be released by the nuclear reactor, could be thrown off badly by such chemical wear.

Thus the nuclear rocket has three major characteristics which we must weigh against one another in assessing the role of the Rover atomic engine for space propulsion. First, it is lighter for a given thrust than a chemical engine and thus will give a larger payload as long as this is unmanned and not sensitive to gamma and neutron irradiation. Second, although the nuclear fuel could be made essentially inexhaustible, the necessity of carrying inert matter for ejection means that the Rover rocket is basically similar to the chemical engines in its one-shot nature. It is good for just one shot and then it must be jettisoned as useless. Third, the intense radiation emitted during take-off, and the radioactivity produced then, mean that the use of the Rover engine is hazardous. It requires careful shielding and isolated launching sites.

For all these reasons we should ask, "Isn't there a better way of pushing rockets with atomic energy?" "Isn't there some way by which we can avoid this problem of carrying hundreds of tons of hydrogen in order that something be available to be thrown away?" "Isn't there a way which is less clumsy and hazardous?"

Well, this may be. One suggestion is to eject the matter at high velocity. As stated earlier, the net thrust on the rocket is proportional to the rate of throwing away matter, that is, pounds per minute, multiplied by the speed it has when thrown, so if the speed can be increased, the amount thrown away can be correspondingly decreased. A possibility here is to give up heat engines entirely and move to speeds which are higher than those which can be obtained from any flame or reactor. The average speed of hydrogen atoms at 8000°F—about the hottest conceivable temperature for reactor devices—is 3 miles per second. This is somewhat less than that of the projectiles from certain types of guns of special design, and obviously, in the face of the fact that in the laboratory many devices exist which can produce atoms at speeds exceeding this by as much as 60,000-fold, one wonders whether we should not try to put such devices, or similar devices, into rockets so that our effort will be exerted more on the ejection at high speeds rather than on the throwing away of large masses at relatively low speeds. If it were possible to use a speed of 1,500 miles per second, for example, a 1 million pound thrust engine would require not 2 or 3 tons, but perhaps a thousandfold less—or about 4 pounds per second. This amount is so small that the nuclear engine now would be unique, coupled with the assurance of being able to fire up again after take-off, and landing, say, on the moon. Perhaps this process of landing and take-off could be repeated several times over. Obviously such an engine has extremely attractive characteristics. So, why isn't it flying, and why isn't it as far along in our plans as Rover?

The answer is twofold. First, the device for accelerating the atoms, the atomic gun, you might well call it, has not been perfected yet. In September, 1958, at the Second Atoms for Peace Conference in Geneva, Switzerland, Dr. John Marshall of the Los Alamos Scientific Laboratory demonstrated a gun of this sort which worked well and reliably for over 2 weeks. His demonstration consisted of making a little metal tab about the size of a dime suspended on a string swing back and recoil when it was struck with a mass of gas atoms (actually, it was ionized gas called plasma) weighing much less than one-millionth as much as the little tab. This was a very striking demonstration, showing that accelerated plasma can carry momentum and can do things much like a bullet or a freight car or other solid object. It was a striking demonstration, but it also was apparently quite far from practicality. However, the principle of this type of engine might be applied to the problem of throwing away the hydrogen, or whatever other gas is used, at speeds which are much greater than those which can be obtained from a chemical fire or from a nuclear reactor. It would then be the plan to use the nuclear reactor not to make heat, but to make electricity which would run the ejector gun.

Unfortunately, there is another weakness in this high-velocity ejector scheme, and in some ways it is more serious than the fact that we haven't yet thoroughly completed the invention of the atomic gun. This is that in increasing the speed with which the matter is thrown away in order to reduce the mass being ejected, the energy used up in the ejection has risen proportionally to the increased speed. So in using a speed which is several times that of the original thermal device, the energy consumption, or the power required for this work, has gone up several times. In other words, a tenfold increase in speed requires a tenfold increase in the power of the reactor needed to furnish the electricity for the atomic gun. Going to the ultimate limit of a factor of 60 thousandfold in speed, the reactor would become so powerful as to begin to resemble things that we have not yet envisaged. Thus the possibility of actually achieving the highest speeds of ejection from a nuclear rocket, or any other kind of rocket, seems to be remote, at least from presently known designs and technology.

One should not, however, lose sight of the fact that improvements probably can be made so that even with the weight of shielding which would be necessary for manned nuclear rockets—and manned rockets certainly are the kind of thing we are thinking about in connection with space flight requiring nuclear engines—a practical nuclear unit can be achieved. It does seem, on weighing the difficulties of the ion gun and the very powerful reactor needed, that the Rover engine, with all of its other troubles, probably is far more practical as of the present moment.

Perhaps some intermediate device which utilizes the Rover engine principle plus the ion gun principle will come into being. A number of technological developments might change this situation rapidly. For example, suppose it were possible to store the electrical energy in lightweight condensers (as yet uninvented), to store it on the ground by the use of ordinary reactors working over periods of weeks and months, for expenditure during the blast-off period of flight, or for other purposes during subsequent periods of flight. Then the problems of weight and radioactivity, which have been bothering us so much in consideration of the Rover engine and the ion gun engine, would be greatly reduced, and we would have left only the development problems of the ion gun. Of course, it might well turn out that this would not be atomic power, for the electricity used to charge the condensers might come from conventional power stations which could well be nonatomic. Anyhow, some day most conventional power stations will be atomic, probably, and so in this broad and long-term sense we do speak of atomic energy in mentioning the lightweight condensers as a possibility in the future development of the ion gun rocket. It would seem that work should be done to utilize the high thrust of the fast throw-away, even though it requires more power, in order to increase the payload, particularly if a lightweight condenser could be developed. This might be the shortest way toward large payloads and practical interplanetary space travel.

Another scheme for avoiding the matter of the slow throw-away in the rocket, and the difficulties of the Rover type engine, was proposed by Stanley Ulam, of the Los Alamos Scientific Laboratory, years ago. Since that time in 1947 some serious work has been done on the basic idea. Dr. Ulam's idea was to produce a thrust for a large space vehicle by using a series of small atomic bombs exploding in rapid succession. That is, atomic bombs would be exploded below a heavy object, possibly a base-plate type of structure, which was thick enough to shield the occupants and apparatus on the top side, and allow the radiation and blast from the nuclear explosions to push on it and produce a useful change in momentum. A type of kick or recoil is received by any solid object near a nuclear detonation, even in a vacuum, and it is calculated that this effect can be used to lift and propel large objects weighing many hundreds of tons. This feature of nuclear explosions has attracted some very able people to work on the idea at the General Atomics Laboratory at La Jolla, and the project in general is very much alive, although everyone considers further development to be difficult in the absence of experimentation. The kind of experimentation required may not be possible so long as nuclear test fall-out is to be strictly forbidden, since the propelling bombs certainly would have to produce some. And it is difficult to see how problems of shock, shielding, and the like could be solved (though we do see that they probably do have solutions) without actually doing field tests to prove that the solutions envisaged are indeed correct. Therefore, this approach is pretty well stopped unless nuclear explosions can be made in the atmosphere, and at the present time this seems to be an unlikely development.

In summary then, the Rover engine is the best we can do at the moment in the way of atomic energy for propulsion of space vehicles, and with all of its difficulties, it may continue at the head of the atomic list for years. In principle, however, high-velocity ejections using the intrinsic properties of atomic energy, which means large amounts of energy in a small mass, should win in the end. It is difficult to judge

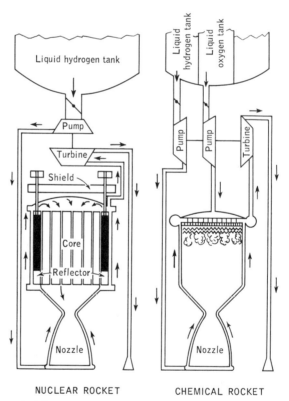

NUCLEAR ROCKET   CHEMICAL ROCKET

FIGURE 41. Schematic drawings of nuclear rocket and chemical rocket.

the practicality of the use of atomic explosions for propulsion without field tests involving attempts to propel heavy objects by nuclear blasts.

Figure 41 is a schematic drawing of the Rover engine alongside an ordinary chemical engine just to show the principles by which the nuclear rocket engine works. Figure 42 shows the first test engine on the test stand at the Nevada Proving Grounds in 1959 just before being fired in the first try. Figure 43 shows the engine en route on its freight car on the railroad track leading to the final test site, and Figure 44 shows the engine actually being fired with the hydrogen gas coming out

FIGURE 42. First nuclear rocket test engine (Kiwi-A).

and burning in the air. This test, which was carried out in June, 1959, was very successful, with very little radioactivity being released, and a new series was scheduled for 1960.

The 1960 tests will involve two reactors in what is being called the Kiwi-A Prime series of tests. (The kiwi is a bird which has the characteristic of not being able to fly.) The Kiwi-A was the name of the first test in 1959 and those in 1960 are called Kiwi-A Prime and Kiwi-A 3. All use the same test cell and other facilities. The original Kiwi-A has been reassembled and hooked up to the stand at Nevada without nuclear parts in it and will be used as a mock-up in cold flow experiments in preparation for the 1960 tests. The assembly of the Kiwi-A Prime reactor is now in progress at Los Alamos where it will undergo cold critical tests, that is, tests of the nuclear reactor to see whether it will actually produce power, before being moved to Nevada. Power operation should begin by mid-1960. Tests of the Kiwi-A 3 reactor will begin shortly after these are completed. The two reactors are being used in order to perform a wide variety of tests and obtain more data than would be possible with one. To determine safety factors and limits of performance, the Kiwi-A 3 tests will include endurance runs until there is positive evidence of damage to the reactor. This, of course, will provide invaluable experimental information. After the runs are completed each reactor will be disassembled and examined. Additionally, there is

FIGURE 43. Kiwi-A test engine en route to final test position.

Figure 44. Kiwi-A test in June, 1959.

now work going on in Nevada to build a new test cell to be ready for 1961. This cell will be similar in design to the existing cell, but will permit more extensive exploratory testing of complete rocket engines. Work on communications and utility facilities has begun, and contracts have been let for the railroad, the vehicle road, water, and various major supporting facilities.

Atomic Power Plants for Satellite and Interplanetary Space Ships

While atomic energy for space propulsion has many difficulties ahead of it before becoming practical, the use of atomic power supplies for space ships or orbiting satellites and interplanetary probes is almost a certainty and has essentially arrived. The problems are not gigantic and they are solvable—in fact, in several instances they are well on their way to solution. We are developing nuclear energy sources which are extremely lightweight compact units and which can provide electrical power for long periods of time. There are two types of these electrical generating units, one using radioactive isotopes and the other using nuclear reactors.

The whole program on the application of atomic energy for supplying nonpropulsive power for space vehicles has the code name of

"SNAP," short for "Systems for Nuclear Auxiliary Power." The SNAP program includes both the isotope avenue and the reactor avenue. It turns out that the isotope avenue yields power sources from a few watts to several hundred watts or possibly a kilowatt of electricity, while the reactors can yield from a few hundred watts on up to many kilowatts, depending on the requirements. For example, the SNAP 8 Reactor now being worked on by Atomics International at Canoga Park envisages the generation of 60 kilowatts of electricity with a total weight of some 3,000 pounds, or about 50 pounds per kilowatt. If we look at space power requirements as shown in Figure 45, we see that they range from a few watts up to 10,000 kilowatts, the lower requirements being for instrumentation and data transmission, and the higher ones being for the electric propulsion mentioned previously. So we see there is plenty of work to be done, and it looks like all possible sources of energy will be found useful in one kind of application or another.

Early in the program it was recognized that the development of the radioisotope power source would take a relatively short time and as a result could provide an early capability. In January of 1959 Mr. John McCone, Chairman of the Atomic Energy Commission, reported to

SPACE POWER REQUIREMENTS

Instrumentation	
Scientific measurements	0.05 - 5 kw
Reconnaissance	0.1 - 10 kw
Environmental control for manned vehicles	1.0 - 50 kw
Communications	
Data transmission	0.01 - 10 kw
Point-to-point radio communications	0.1 - 10 kw
World-wide radio and TV broadcasting	0.1 - 1,000 kw
Jamming	1.0 - 10,000 kw
Electric propulsion	
Maneuverable satellites	30 - 1,000 kw
Lunar and interplanetary probes	100 - 10,000 kw

FIGURE 45

the President and announced publicly the development of a 5-pound device, 4¾ inches in diameter and 5½ inches high, which generated about 5 watts of electricity from the radiation emitted from one-third of a gram of a radioactive isotope, in this instance polonium 210. This device, called SNAP 3, was a "proof of principle" effort developed in practical size and with a practical power rating. It was developed by the Martin Company in Baltimore, Maryland, working with the Minnesota Mining and Manufacturing Company, the latter having developed the thermocouples, consisting of a compound of lead telluride alloyed with other substances, which convert the heat of the nuclear reaction directly into electricity with no moving parts. It is obvious that SNAP 3 could have been powered by many other radioactive isotopes, but polonium 210 was chosen because it was handy. SNAP 3 is expected to produce electricity at the rate of 3 watts and have an over-all efficiency of 5 or 6 per cent. Its advantages are its very small size, light weight, long life, and reliability. There is nothing to break in it and nothing to go wrong. Now, over a year later, we can say that the early polonium 210 version did operate and did fulfill the expectations. In SNAP 3 (Figure 46) there are 20 pairs of electricity-producing thermocouples stuck around the spherical isotope source contained safely in the metal sphere. It generates about 3 to 6 watts, depending on the activity of the isotope.

It is interesting to notice that the SNAP 3 device need not necessarily be utilized in outer space only. In fact, any remotely operated equipment requiring 3 to 5 watts of electric power is a potential user. And in some instances we find that the cold junction can be kept colder than in the case for space flight, resulting in generating more electric power. This would be true for most applications on the surface of the earth. Some of these might be remote weather stations, airlane markers and warning beacons, navigational buoys for sea lanes, sonobuoys for submarine detection, and so on. A particularly attractive feature of SNAP 3 is the safety with which the isotope can be contained, so in case a satellite carrying it falls to the ground there will be no dissemination of radioactivity. There is no reason why the SNAP 3 sphere cannot be so tightly sealed that it can be used as a gun projectile and still not burst and scatter radioisotopes. In fact, such rigorous tests have been made, looking forward to the utilization of such isotopes as strontium 90 for this purpose. It seems entirely possible to use it safely as a satellite power source and also as a power source for remote installations here on earth. The amount of isotopes required is rather large, and therefore it is only the cheaper of the isotopes that make the thing practical. Take strontium 90, for example; the thermal power available is about one-third of a kilowatt per pound of the isotope mixture. Our present annual production of strontium 90, considering everything, in the United States would not amount to more than a few kilowatts of electricity on this basis, so it seems conceivable that satellite power alone

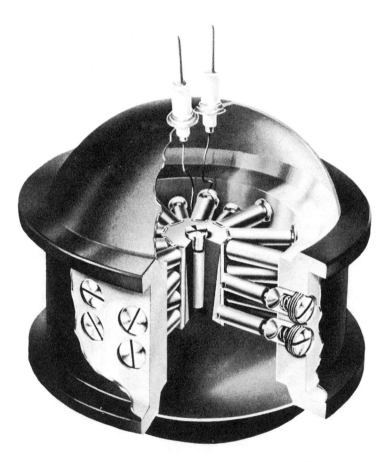

FIGURE 46. Isotope power source for satellites (SNAP 3).

may constitute a major outlet for this isotope. Other, shorter-lived isotopes may be preferred for reasons of higher power for shorter times, and there are several of these; but in general the fact is that our production capability is not vastly greater than the amount needed for the SNAP 3 type of satellite power source and the uses it can be seen to have in the near future. This, however, is not a disadvantage at this time for no other uses have as yet been developed which are commensurate with the SNAP use in the quantities of isotopes required. The other uses involve the radioactive radiations themselves and do not use these radiations merely to make heat; hence, by comparison they require insignificant quantities of isotope.

In SNAP 3, however, we do have a power source which is certain to work, and which will work in the case of strontium 90 for 40 years on the average (providing the thermoelectric elements hold up for this time, which they probably would). So a satellite that is launched with a strontium-90 power source would probably continue to operate, as far as power is concerned, for many, many years. And to orient ourselves on the power requirements, Pioneer V has talked to us from 22 million miles with only 5 watts of power, well within the strontium-90 SNAP 3 capabilities.

Now we consider nuclear energy for space application simply because the nuclear process offers the most energy per unit of mass of any source known, but there are logistic problems, which I mentioned in the case of propulsion, which cause real complications in the use of nuclear energy. In the matter of power sources of a few hundred watts, or even as much as a kilowatt, the solar cell power source—that is, the collection device itself plus rechargeable batteries which are then used to operate equipment—is a real competitor. An announcement in 1960 on the Thompson Ramo Wooldridge contract to develop a solar power source, describes a device with a collection area of some 1,000 square feet which would unfold after launching and would develop something like 3 kilowatts of electricity in a total weight of 700 pounds. This is a very impressive competitor for the radioisotope source and, in fact, is pushing on the nuclear reactor power source. One problem, of course, is that the batteries weigh something, and another is that sunlight is not always present. There are many areas which need exploring—for example, the back side of the moon—which in certain periods would not have sunlight, so instruments operating in the nighttime could not be powered by solar cells alone; they would need batteries, and batteries are notoriously heavy. (We might at this point note again the earlier remarks about the importance of the development of lightweight condensers. This would really help in the solar power source as well.) There are other similar space missions where solar units will probably not be able to perform satisfactorily, such as in the shadows of the planets and on the down-sun side of the solar system where the sunlight intensity diminishes rapidly, or in a flight to Venus where the satellite would go under the clouds of Venus. These are all special instances where the SNAP radioisotope power source is uniquely qualified and essential.

As stated previously, in the range of a few hundred watts to tens and hundreds of kilowatts, atomic power comes only from nuclear reactors. The initial effort on nuclear reactors in the SNAP program was SNAP 2, a reactor-powered turboelectric generator for space vehicles. This reactor consists of a bundle of cylinders, each of which is 1 inch in diameter and contains a 10-inch long fuel element using uranium 235. It is not pure uranium 235 but is mixed with zirconium, and the two metals are treated with hydrogen to make a mixture of zirconium and uranium hydride. There are 61 of these stainless steel cylinders and at the end of each cylinder there are short 1½-inch plugs of beryllium to reflect the neutrons. This whole bundle fits into a can about 8 inches across and 14 inches tall as shown in Figure 47. The active portion, that is, the heat source of this reactor, has a total volume slightly under ⅓ cubic foot. The control of the nuclear reaction is accomplished by means of rotating drums made of proper materials asymmetric in composition around the axis of rotation. The first model of this SNAP 2 reactor achieved criticality in September, 1959, and was brought to full power and temperature the following November. This was done at Canoga Park by Atomics International, under contract with the Atomic Energy Commission.

The objective of the SNAP 2 project is to develop a space power system for use in the period 1962 to 1963. It is supposed to operate continuously without interruption for 1 year at a power level of 3 kilowatts. The gross weight including the radiator is to be 620 pounds. It is to operate without maintenance and is not to start up until it is in orbit, so it will constitute a negligible hazard during launching because it will not have been activated. The reactor operates with sodium-potassium alloy taking the heat and passing it on to a mercury boiler, the mercury vapor from which runs a turbine which extracts the energy and converts it into electricity in the attached alternator. These little generators and turbines are the tiniest things, but they have been proved to operate reliably. The nuclear fuel charge involves about 53 kilograms of uranium 235 and the total weight of the reactor is 220 pounds. The components of the power conversion system were developed by Thompson Ramo Wooldridge. There is little doubt that the SNAP 2 reactor system (Figure 48) will attain the objectives set for it.

Now what about the future? I think in speaking of the future it is well to be aware of the plans for SNAP 8. The National Aeronautics and Space Administration and the AEC have agreed to jointly develop the SNAP 8, with the idea of using it as a source of electrical power for the high-velocity throw-away I have been mentioning, that is, propulsion by ions. The AEC will take the job of building the reactor and NASA will provide the machinery for making the electric power. It probably will be very similar to the SNAP 2 in using mercury, the

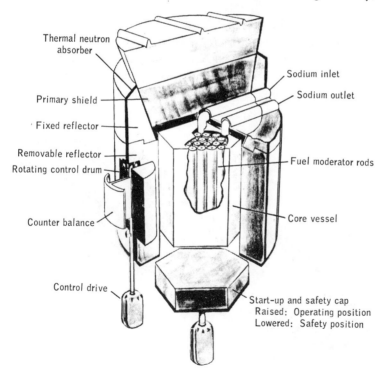

FIGURE 47. Reactor power source for satellites (SNAP 2).

vapor of which will be passed through a turbine. The difference will
be that the electric power output of the generator will be of the order
of 30 kilowatts, ten times that of SNAP 2. Actually, the reactor will be
large enough to produce 60 kilowatts of electric power, and so the
generator will probably have a capability of 60 kilowatts. These two
power levels will give greater flexibility in the application of SNAP 8
than just a single power level would. The 60 kilowatt model will weigh
about 3,000 pounds, or 50 pounds per kilowatt.

It is assumed that this 60 kilowatt version of SNAP 8 will be launched
into an orbit that is about 300 miles above the earth, and the complete
satellite in this orbit we may assume would weigh about 9,000 pounds,
which would be about the capacity of the Centaur launching vehicle.
After stable orbit around the earth has been attained, the reactor and
the atomic gun machinery would be started. This would gradually lift
the satellite into higher and higher orbits, and eventually it might
reach an orbit 22,000 miles above the earth, which is the 24-hour orbit.
This is the orbit where the satellite stands still above any given spot on
the ground. It is estimated that in 40 days SNAP 8 could raise a net
payload of 3,000 pounds to this orbit. This is according to the testimony
of Mr. Robert E. English of the NASA Lewis Research Center, before
the Joint Committee on Atomic Energy of the Congress. In contrast, a
high-performance chemical rocket can place about a thousand-pound

Electrical power output	3 kw
Reactor power	45 kw
System efficiency	6.7%
System weight	500 lb
Radiator area	85 ft^2
Reactor outlet temperature	1200°F
Radiator temperature	600°F

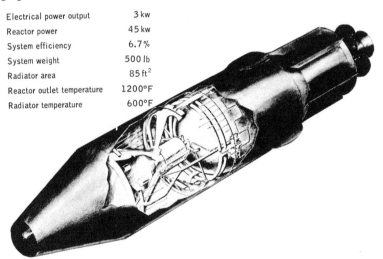

FIGURE 48. Reactor power system for satellites (SNAP 2).

payload into the 24-hour orbit at the equator, and for a communication satellite—and this would be one of the principal reasons for putting something into the 24-hour orbit—part of that thousand pounds would have to be devoted to communications equipment and part to the electric power supply. The significant point about SNAP 8 is that it has power supply in addition to its payload. In other words, the 60 kilowatt reactor is used for running the throw-away gun during propulsion, but after the satellite is in orbit it can be used for broadcasting or other purposes.

In addition to electric propulsion demonstrations and planetary probe auxiliary power, the SNAP family of nuclear power plants can do remarkable things in the way of utilization of space for peaceful purposes within this decade. This type of device, taken together with the Centaur and Saturn launching vehicles, could provide this country with several global communication systems during the latter half of this decade. For instance, we could have low-cost, all-weather, intercontinental telephone systems, although there is an interesting point about these telephones. Since the distance to the satellite relay and back is at least 44,000 miles, even at the velocity of light this involves a delay of about ¼ second in transmission, so we would have to learn to talk this way. It isn't difficult, but it is noticeably different. There could be many channels of FM radio broadcasting coverage that could be received in all parts of the world, regardless of the degree of local economic development, and it seems likely that the SNAP 8 system could provide sufficient power for a channel of continuous television broadcasting on a completely global basis. World-wide multichannel all-weather television broadcasting from three fixed 24-hour-orbit satellites could reach all the villages and all the peoples of the world equipped with low-cost television receivers. It could also provide sufficient electric power for near-earth satellites to accomplish precise air traffic control and navigational aid functions, such as the first of the navigation satellites, launched in April, 1960. These would be major accomplishments in the peacetime use of space—accomplishments for which the atom and atomic energy are necessary, as we see it now.

Conclusion

In the future, when the question of making trips to distant places in the planetary system is a realistic one, the role of atomic energy will certainly grow. We must continually bear in mind the matter of its lethal radiation and the need for shielding of manned vehicles, although with radioactive isotope power sources in strengths up to several tens of watts, this is not a major problem. Also, in actually orbiting satellites and interplanetary space ships it is not as serious a problem as one might imagine, for there is no air to back-scatter the radiation. One of the serious problems in designing reactor shielding here on earth is that the air scatters radiation back, and hence around the shield. Otherwise, a simple wall between the cabin and the reactor itself would be adequate. Nevertheless, we must always have a certain weight devoted to shielding when we use atomic reactors, and this is a negative characteristic—one which is against the use of atomic energy in space vehicles.

On the other hand, the many positive characteristics make it absolutely clear that the role of atomic energy in space is to be a significant one. And we must hope that the Atomic Energy Commission and the National Aeronautics and Space Administration continue to cooperate wholeheartedly to try to develop jointly the potentialities of these great new areas. Certainly there is little that is more exciting in the world today than some of the peacetime uses of space. The world-wide television system that SNAP 8 would make possible is hard to overestimate in importance and in its potential effects on our lives, for with it would come all the other capabilities for rapid communication, navigational aids, and the like.

It is essential that in this new period we all become students and try to learn about space and atomic energy. In many senses the rate of our exploitation of these new areas will be determined very largely by the rate at which we learn about them. Personally, many years spent in the atomic energy program have schooled me in this area, but I find it necessary to spend considerable time in keeping up with the new space developments and in thinking about the new problems and the areas where atomic energy might be applied to space. And I am convinced that there are many important ones which have not yet been either thought of or mentioned. Therefore, there is a large payoff in educating more technical people in these areas. The Atomic Energy Commission has had for many years an educational program of training engineers and other students in atomic energy. I would strongly recommend that NASA follow a similar program and take on educational activities among scientists and engineers as one of its major long-range responsibilities. This, in the end, is the only way that NASA can do its job and that the country can exploit and develop to the utmost the capabilities in these great new fields.

The Atomic Energy Commission and NASA, working together closely in this effort, can take the leadership of the world in the space program for our country and keep it. There is no doubt that we are in many areas pre-eminent, but there is also no doubt that we are not in others. Far more important than the question of a million pound thrust rocket engine is the question of whether the next generation of engineers and scientists will be interested in space. Because the Atomic Energy Commission successfully set up an educational system, atomic energy engineers are now in good supply. Scientists are pretty well educated in atomic energy. A similar program should be—it must be—the objective of NASA in order that the maximum rate of progress be made. I am certain all of us in the universities of this country would cooperate with them to the utmost in the development of such a program. It is clear to educators that this is a necessity; it is clear to administrators that this is a necessity; and being both an ex-administrator and, I hope, an educator, I see this particularly clearly and want to make it the concluding point of my remarks.

We must see to it that young scientists and engineers think about and study SPACE. Space and the future are not synonymous, but they are certainly closely linked.

9

The Place of Government
in the Utilization of Space

OVERTON BROOKS

CHAIRMAN, COMMITTEE ON SCIENCE AND ASTRONAUTICS
HOUSE OF REPRESENTATIVES, UNITED STATES CONGRESS

LL.B., Louisiana State University, 1923; at which time he began practicing law at Shreveport, Louisiana; United States Commissioner 1925 to 1935. Elected to the Seventy-fifth and to the eleven succeeding Congresses and is now serving in his twenty-third year. Chairman of the Board of the National Rivers and Harbors Congress, to which position he has recently been elevated, having served for 5 years as President. Vice Chairman of the House Committee on Armed Services for many years, giving up this position when he became Chairman of the House Committee on Science and Astronautics, which position he now holds. Member of the Board of Regents of the Smithsonian Institution.

MY TOPIC is the place of government in the utilization of space. In dealing with so broad a subject matter, it will be necessary to pick out certain aspects as most suitable to the purposes of this book.

I will begin with the present circumstances of the national space program, the values we can expect it to achieve, its importance to our international position, proposed changes, and historical background.

Then I will take up some problems of space regulation and control. Space law is not only needed but needed urgently, in view of the facts of space and the participation by other countries in space exploration and research. Later, it will also be necessary to regulate the activities of private enterprise in space. Up to now, all American space activities have been conducted by the national government. For some time to come, the government will continue to be the principal space entrepreneur. In a society dedicated to private enterprise, however, it seems inevitable that space will sooner or later be used for private purposes. Space law must anticipate this development.

Finally, I will list some steps that would enable this country to reap the benefits of space technology both sooner and more certainly. Time is short. The opportunities are vast.

Need and Value of Astronautics

The year 1960 began in Washington with a great debate over the national space program. I wish I could tell you that the outcome was a clear set of conclusions. Unhappily, the charges and rebuttals are still flying through the air like ballistic missiles. Many outsiders complain that they feel confused. Some insiders may secretly feel the same way, though they may never admit it. And no wonder! For example, at least four different sets of comparisons of American and Soviet missile strength have appeared in major newspapers of the nation.

I am not going to try to clear up the missile gap or even the space lag. The roots of the disagreement go deeper. If you want to know about the prospects and problems of using outer space, including our chances of outstripping the U.S.S.R. in the long run, and what our government can do to help, you need information of a more fundamental kind. Perhaps we should return to fundamentals and start with the need and value of space flight. The debate in Washington has revealed a difference of opinion on this subject, even within the executive branch of the government.

People have been dreaming about space travel for thousands of years. Now that we have the means to make the dream come true, however, many people begin to doubt the value of our space program. Is it merely political and psychological—that is, are we going into outer space just for reasons of national prestige and advantage in the cold war? Or is it military—to prevent outer space from being used against us, and to use it, if need be, against our enemies? I would answer that the value of our space program is both psychological and military, in the cold war and the hot war alike, and could be amply justified on either ground, in the perspective of 10 or 20 years. Or we can justify it on the ground that it leads mankind farther along the roadway toward his destiny.

More directly, for most of us, the value of our space program is also scientific and economic. If you happen to be a teacher or a housewife, it may be hard to think of outer space as something that affects you. Yet it does just this; and sooner or later much of our prosperity as well as our national security is going to be wrapped up in outer space.

For example, we can look forward to better and cheaper methods of radio communication, weather forecasting, and navigation by means of satellites. Already new methods of treating diseases and of analyzing the human anatomy, which will have a direct effect upon human beings, are being developed in our space program. New space exploration may unlock the ancient secrets about the origin and use of the solar system and even of life itself.

Admittedly, such wonders lie in the future. Yet I expect to see some of them myself in the next 10 years or so.

At present, space exploration is first and foremost a new means of scientific research. Naturally, this aspect appeals to the scientific community, both at home and abroad. It is only the beginning.

Military and Civilian Uses of Space

From what I have just said, you can see that the benefits we hope to gain in outer space are both military and civilian. We all welcome the fact that the value of space exploration is apparently much greater than just military possibilities. At the same time, the relationship between the military and civilian space programs has raised problems from the very beginning. The first successful American satellite was launched with a military missile in the hands of military personnel. We thus began to acquire the ability to explore space, in the first place, through military efforts to develop missiles as weapons of war. The U.S.S.R. had the same experience.

All concerned, including our military leaders, agree that space exploration is not solely a military program. Neither can we deny that space has tremendous military significance. The problem, then, is either to separate military and civilian space functions, or to link them in some effective relationship.

As many of you know, this kind of problem is not confined to space alone. It is a perennial problem of government in dealing with new technology. For example, the Atomic Energy Commission was given the job of providing nuclear warheads for military missiles. The armed services, however, remained responsible for the means of delivery and for developing and using the military weapon systems. In the field of atomic energy, then, civilian and military functions are not entirely separated. This arrangement has worked well.

The Space Act of 1958 created a new civilian space agency. The result is a system of coordinate civilian-military control, which was recommended by President Eisenhower in his message to Congress of April 2, 1958. In its final form the act had the full approval and signature of the President.

The civilian space agency was given general authority to plan, direct, and conduct aeronautical and space activities. At the same time, the act provided for full cooperation between the civilian agency and the defense establishment. Because civilian and military interests in space are often difficult to separate and unavoidably overlap, machinery was created both for coordination and for over-all control. For this purpose the act established a National Aeronautics and Space Council, of which the President has been Chairman, and a Civilian-Military Liaison Committee for the day-to-day coordination of civilian and military space activities.

As time has gone on, it has become apparent that neither the Space Council nor the Liaison Committee was used effectively. On the other hand, there is general agreement that the Space Act provides an adequate basic framework for national space activities.

In a message to Congress, President Eisenhower declared that ". . . a single program embracing military as well as nonmilitary space activities . . . does not exist and is in fact unattainable."

There are certain questions that trouble me on this score: Are the military and civilian space programs really so separate and distinct? How should they be coordinated? How should the space effort be divided between military and civilian agencies? Is there a need for long-term planning, direction, and coordination, encompassing both the military and the civilian space programs, and, if so, how should the need be met?

On all these questions there is a clear-cut difference of opinion. For example, in contrast to President Eisenhower's views, General Medaris told the Committee on Science and Astronautics:

> I believe strongly, and feel that it is wholly demonstrable, that the fields of ballistic missilery and space exploration and exploitation are in fact naturally indivisible elements of a single broad technology and that a continuance of divided efforts in this broad area cannot but result in delay, duplication, and waste of both money and manpower. . . .
>
> From a purely technical viewpoint, there is so little difference between civilian and military space programs that there is no justification for their division and resulting duplication.

President Eisenhower indicated that in his mind this government is not engaged in a space race with the U.S.S.R. For example, he was asked whether he felt any sense of urgency in catching up with Russia either in space exploration or in military missiles. He replied (according to the New York Times): "I am always a little bit amazed at this business of catching up. What you want is enough, a thing that is adequate." We may naturally ask: "Adequate for what?" Surely not adequate for catching up with Russia.

Again, when he was asked whether our national prestige was at stake in space exploration, he replied: "Not particularly, no." President Eisenhower also said in his 1960 State of the Union Message: ". . . our effort in space exploration . . . is often mistakenly supposed to be an integral part of defense research and development."

This fits in well with the other statements I have quoted. The view that we are not or need not be in a space race with the U.S.S.R., in my opinion, ignores both the effect of space achievements on other nations and the military potential of space technology—which I believe cannot be denied.

Status and Prospects of the National Space Program

At present, the space program is drifting with the winds of expediency. While continuing to deny that our space program involved national prestige, competed with the Soviet program, or required any particular urgency, the Eisenhower Administration nevertheless kept devoting more money to the development of this program. Federal funding of the space program must of necessity increase year by year. The demands of NASA will increase, and the space development of the armed services will not diminish for many years.

Only three Polaris submarines were included in the current budget. Without prior warning, Admiral Burke stated on February 8, 1960, that the Navy would ask for funds to build six more.

The Eisenhower Administration directed a drastic cut in funds requested by the Atomic Energy Commission to develop a nuclear-powered space rocket (Project Rover). As a result, the project was expected to be delayed by a year or more. There are those who say that the delay was not justified on technical grounds. On March 8, 1960, the AEC transferred funds within its own budget to add 11 million dollars to Project Rover.

Naturally, I am pleased that these vital programs will be getting more support. It does seem to me, though, that they should have had adequate support at an earlier date in accordance with an over-all plan. Must we proceed by fits and starts—one step forward and two steps back? Surely it is costly and dangerous to play our national space policy by ear.

Suppose we take a longer look ahead. What are the long-term plans for our space program?

On January 28, 1960, the National Aeronautics and Space Administration gave the Committee on Science and Astronautics an estimate of its budget needs for the next 10 years. The total was less than 15 billion dollars, or an average annual rate of some 1½ billion. Recall that the agency received more than 500 million dollars for the fiscal year 1960—its first full year of operation—and requested 915 million dollars for the fiscal year 1961.

Yet the agency presented its 10-year budget forecast as sufficient to wrest space leadership from the U.S.S.R.

Although it seems to lack sufficient urgency, the 10-year program is technically well thought out. It called for the flight of a Mercury astronaut (not in orbit) in 1960; an unmanned "hard" landing on the moon in 1961; planetary probes of Mars and Venus in 1962; and an unmanned "soft" landing on the moon in 1963.

From the standard of meeting our competition, however, it seems too low, overoptimistic, and unrealistic.

For example, the only big space boosters included in the 10-year program are those already under research and development: the Saturn and the Nova. As you know, our space lag is mainly in boosters. The Saturn is expected to become available by 1965. Four F-1 engines (six in some versions), each producing 1½ million pounds of thrust, will be clustered to make the Nova. The first flight test of the F-1 engine is not scheduled until 1968. No date has been set for completion of the Nova. If the U.S.S.R. develops even bigger boosters in the meantime— before 1970—our space budget will need to be raised very sharply, or we will fail to reach the goal of catching up. Dr. Wernher von Braun told my committee: "I consider it quite likely the Russians have a larger rocket than any they have flown so far."

To give another example, manned flight to the moon is not scheduled in the 10-year program until *after* 1970. Dr. von Braun also told my committee that he "would not be surprised if Russia makes a soft landing on the moon this year."

Can we catch up with the Soviet Union? I believe we can, and I have yet to meet any American who thinks otherwise. Even Khrushchev seems to agree. At the U.S. Fair in Moscow in 1959, Khrushchev talked with Dr. John Turkevich, a Princeton professor who speaks fluent Russian. Dr. Turkevich later reported that he said to Khrushchev: "We will catch up to you and beat you in satellite weight before you surpass us in wheat production!" (An astute remark in more ways than one.) Khrushchev replied seriously: "America is a strong and powerful country. If it sets its mind to do something, it will succeed."

The Space Lag

Ever since the first Russian satellite was launched in 1957, we have heard loud cries that we were lagging behind the Russians in space. This charge is not to be lightly brushed aside. Qualified witnesses have agreed that the main-stage rocket, or booster, used in Soviet space missions produces from 600,000 to 800,000 pounds of thrust. In this country, the largest rocket booster is still the 360,000-pound-thrust Atlas. We are trying to bridge the gap by developing high-energy upper stages as well as bigger boosters. Within a few years, we may have perfected the giant Saturn rocket. In the meantime, however, the U.S.S.R. is not likely to stand still.

According to testimony given to my committee, the Soviet space lead is based primarily on this two-to-one superiority in rocket thrust. Greater thrust permits the U.S.S.R. to carry out space missions that would otherwise be impossible. It also shortens lead-time and increases reliability.

As you all know, the result has been a string of "firsts" and other exhibitions of Soviet prowess in rocketry. These are not hollow triumphs; they reflect an ability to put men and instruments into orbit, or into deep space, for a variety of purposes. They have also convinced many people around the world of Soviet leadership, not only in rocketry but in broad areas of science and technology. While it is clearly a delusion to take rocketry alone as a proof of general scientific progress, the fact that many people do so has affected the world position of the United States.

The impact of Soviet space feats on world opinion was finally recognized by many high officials of the Eisenhower Administration—including Dr. George Kistiakowsky, the President's scientific adviser, who said that we ". . . cannot ignore the very real political implications of various spectacular accomplishments in outer space that have come to have symbolic meaning to the world at large."

A month or so earlier, Dr. Kistiakowsky observed that this country is engaged in "a scientific and technological contest with the Soviet Union, which today involves our national prestige, and tomorrow, perhaps, our very survival."

Now, I am not recommending national self-flagellation, though I think a little self-criticism is good for the soul and good for the country. On the contrary, I believe that Soviet superiority in rocket thrust, while undeniably an important index of space capabilities, should be kept in proper perspective.

Whether in space research or military astronautics, the best results are not always obtained with the biggest boosters. Nor can progress in space research be measured by the chronology of spectacular feats. There are too many partial measures which cannot simply be added up— for instance, the variety and types of missiles and space vehicles, the number of successful launchings, instrumentation, reliability, tracking, and control devices.

Dr. Homer Newell of the National Aeronautics and Space Administration recently completed a study of American and Soviet space research programs. He concluded that the U.S.S.R. leads us in high-altitude measurements, biological experiments, payload package ejection, deep-space probes, and vehicle technology. On the other hand, he said, the two countries appear to be nearly equal in upper-air research and in satellite studies of the earth's environment, while the United States has taken the lead in solar radiation experiments and in studies of the atmosphere up to an altitude of about 200 miles.

The detailed comparisons made by Dr. Newell may be cause for concern but hardly for despair. In fact, he might well have included a list of the very creditable American "firsts" in outer space—for example, putting the first satellite into a polar orbit; putting the first satellite into a (near) circular orbit; the first ejection of capsules from satellites; and the first use of a satellite for communications relay. It is widely believed, furthermore, that we have more than held our own in guidance, miniaturization, and data recovery. There is even some feeling, though it is hard to verify, that we have acquired more useful scientific data.

I particularly wish to point out that our space lag, in comparison with the U.S.S.R., is the result of conscious choices. It is not the result of inability to do things in outer space as well and as fast as the Russians. In fact, the Russians have developed more powerful rockets than ours because they began their development 5 to 6 years earlier.

We did not give high priority to ballistic missiles until 1954, when we first knew that we could make light, small, nuclear warheads. Government witnesses told my committee that we made a choice at that time between lighter warheads and more powerful rockets. We chose the lighter warheads. Why could we not pursue both courses, and make both the lighter warheads and the more powerful rockets? If we had done so, according to expert testimony, we might not now be trailing the Russians in outer space.

Is it possible that the space and missile programs have been tailored to fit budgetary preconceptions?

In August of 1955, Dr. von Braun told a high-level meeting that he had the hardware to put the world's first satellite into orbit. He was not allowed to proceed.

In 1956, General Medaris received an order from Washington, instructing him not to let the Jupiter C go into orbit. At that time, General Medaris and his scientific colleagues were about to test-fire the Jupiter C. They believed it could be put into orbit at will.

Remember, the first Russian satellite was not launched until October, 1957.

Yet, as you know, our Thor and Jupiter IRBMs are now available in quantity, and our Atlas ICBMs became operational at about the same time as their Russian counterparts. Considering the dates when we began our programs, it appears that we developed IRBMs and ICBMs in about half as much time as the Russians.

This is an accomplishment in which we can all take pride. But it was made necessary by our own earlier decisions not to develop large rocket boosters and not to engage in a satellite program. The same decisions have hampered our efforts in outer space up to this very day.

Space Law and International Cooperation

So far, I have been talking about our own national space program, what it means to us, and how it seems to be progressing. Now I should like to turn for a moment to the less evident, but no less vital, problems of space regulation and control.

There is a need now to begin work on the establishment of general space law. Man's ability to explore and use outer space is developing rapidly. Not only the United States and the U.S.S.R. but many other nations will soon be in the game. Any delay in creating applicable rules of international law may make our later efforts more difficult.

The need for space law is more than a mere matter of traffic control, important though that may be. Perhaps it would help to give a few examples.

A pressing problem of space law concerns responsibility for the effects caused by the return of space craft to earth. In addition to the need for international agreements dealing with liability for any resulting injury or damage, it may well be necessary to change domestic law. At present, it is by no means certain that an American citizen harmed by an American space vehicle could recover damages in an action against the government.

Another problem is the allocation of radio frequencies for space use. Frequencies have been used in violation of treaty provisions, and transmissions have continued, interfering with normal communications, long after their purpose had been served. Looking to the future, we can see a coming "war" of the radio-frequency spectrum. In effect, the spectrum is a scarce natural resource, already overloaded, which will be required to carry more and more traffic. The growing overload will come not only from military but from scientific and commercial use. Rapid communications will be needed among an increasing number of points for production, transportation, and other economic activities. In outer space, unless agreement is reached on frequency allocations, the information acquired by difficult and costly experiments may well be lost.

To give another example, scientists are beseeching their governments to help in keeping space clean. At the first international meeting on space science, held at Nice in January, 1960, great concern was expressed over the dangers of contaminating other planets with living organisms from the earth and of contaminating the earth if space vehicles return from other planets. In H. G. Wells' novel, *War of the Worlds*, man himself could not overcome the invaders from Mars, but man's diseases did. The dangers of interplanetary contamination are far-reaching and not at all fanciful. They are certainly a proper subject for space law.

It is apparent that new international accommodations will be necessary in the peaceful uses of outer space. Without trying to make a complete list, we can see now that conflicts may arise, unless an understanding is reached, in a number of fields, including the extension of national sovereignty into outer space, interference among space systems, and rivalry on the moon and the planets.

There is a more positive side to the novel forms of international cooperation which are likely, if not certain, to develop as a result. It is true that the use of outer space may create new international problems and tensions. We should not forget, however, that it also increases the incentives and opportunities for nations to live together in peace and harmony.

I will not trespass any further on the domain of the space lawyers except to repeat that their work and its practical applications are of great concern to us all. In order to benefit from the full fruits of space exploration, I believe, we must proceed at once to develop the principles of space law and try to reach early agreement on the scientific and commercial uses of space.

Conclusions

To me, the most hopeful omen of our future space progress is the growth of public understanding and support. In 1958, there were many doubters. Today, most informed people in this country seem convinced of the necessity for going ahead in space at urgent speed.

Let me suggest some definite steps that would enable us to make the best possible effort. My suggestions will not be startling. In fact, I believe that each of these steps is self-evident and that most of you will find yourselves in agreement that each should be taken.

First, recognize that we are engaged in a long-run space race with the U.S.S.R., and make up our minds to compete.

I am speaking, of course, about the scientific and peaceful investigation of outer space. In the field of military missiles, however, there are two matters that deserve special mention: defense against the ICBM, and the Polaris program.

An adequate means of defense against ICBMs has not yet been developed. We should not rest until it has been perfected.

The Polaris is not just another missile system. It is a different kind of weapon—an almost invulnerable means of inevitable retaliation, which can be made ready in time. To close the missile gap by matching the U.S.S.R. missile for missile would be costly, protracted, and perhaps unprofitable in the end; to leap it with Polaris could transform our whole strategic situation.

Second, emphasize large rocket boosters and high-energy upper stages, including the possibilities of nuclear and ion propulsion.

Third, improve the management of our military and civilian space programs. I will not try to predict what precise form of organization will finally emerge. There is a crying need, however, for comprehensive long-range planning, clear lines of authority, vigorous leadership, and continuity of effort.

Finally, further emphasize basic and supporting research and scientific training—the indispensable reservoir on which all our future progress must depend.

10

Competitive
Private Enterprise in Space

RALPH J. CORDINER

CHAIRMAN OF THE BOARD
GENERAL ELECTRIC COMPANY

Ralph Cordiner is an articulate spokesman for modern business. A native of Walla Walla, Washington, with a B.A. from Whitman College, he has spent practically all of his business career with General Electric. Since he became the company's chief executive officer in 1950, General Electric has expanded rapidly, becoming a leader in such space-oriented technologies as electronics, flight propulsion, high-temperature materials, and power supply. Mr. Cordiner was Vice Chairman of the War Production Board in World War II, and chairman of the "Cordiner Committee" on military manpower. He is author of New Frontiers for Professional Managers, *and was selected in 1960 by the* Saturday Review *as "Businessman of the Year." In 1958, Mr. Cordiner received the Certificate of Appreciation, highest civilian award of the Department of Defense.*

THERE ARE two basic forms of human organization: the free societies, such as the United States; and the regimented societies, such as the Soviet Union. In free societies, the government is the servant of the people. In regimented societies, the people are the servants of the government.

No society is completely free or completely regimented, but the distinction is real, and it is profoundly important to the quality of human life. The two systems are now engaged in a protracted struggle for the future of the world. They are competing in every field of human endeavor, including the field of economic development.

A distinguishing feature of the free societies, as opposed to communist and other socialist systems, is the use of competitive private enterprise as the primary means of economic development. The citizens of the United States have both philosophical and practical reasons for preferring business enterprise to government enterprise. Philosophically, the competitive private-enterprise approach is more appropriate to a free society than government-owned or government-controlled industry, which is one of the characteristic features of a regimented society. And practically speaking, the system of competitive private enterprise has enabled this country to produce a level of living that is unmatched anywhere, anytime.

Even the rulers of Russia and China cannot think of a more effective incentive for their regimented people than the hope that some day they may attain the standards of the people of the United States. Communism and socialism may claim to be the revolutionary systems of our times, but for almost every human being outside the Free World, the most successful imaginable revolution would be one that would enable him to live like the average United States citizen.

Space Raises Questions

Now, the two contending systems are beginning to compete in a new area: in the exploration and economic development of space. This will

214

be a vast and expensive undertaking, and it poses some fundamental questions for the citizens of the United States.

How can we utilize our dynamic system of competitive private enterprise in space, as on earth, to make newly discovered resources useful to man?

How can private enterprise and private capital make their maximum contribution? What projects will necessarily require government chairmanship and support for their execution?

What must be done to preserve a free society while competing in an international race for space? How can we assure that when the space frontier is developed, it will be an area of freedom rather than regimentation?

Specifically, here is my objective in this chapter:

First, I will submit for your consideration some deeply important impacts of the space challenge on this country's free society and its system of competitive private enterprise. Many will be highly beneficial. Others have chilling implications. And some are ambiguous at present. All of them will affect the life of the individual citizen and his children, and are therefore worthy of his informed interest and concern.

Next, we will examine some of the underlying principles of the United States political and economic system—principles that must be recognized in the formulation of national space policy.

On this basis of principles, I will then suggest the specific roles of industry and government in the exploration and economic development of space.

Then finally, I will suggest some changes in the legal framework that would encourage the maximum contribution of competitive private enterprise.

Let us begin by looking at the future impact of space activities on the free society and private enterprise.

There will be many effects that are primarily beneficial.

A New Frontier

The advanced industrial nations can now send objects off the planet into space. This new capability opens a whole new frontier to human exploration, development, and use. At this stage, the new frontier does not look very promising to the profit-minded businessman, or to the tax-minded citizen. But then, so it must have seemed to the Greeks, when Jason returned from his exploratory trip into the Black Sea; or to the Phoenicians, when their first explorers returned from the wild and savage shores of the Western Mediterranean. Most of the Greek and Phoenician traders, in 1000 B.C., probably preferred to invest their money in a good safe cargo of grain or wine, shipped over familiar sea lanes to familiar markets. But apparently a few traders, and later colonists, had the vision to see possibilities where other men saw nothing, because in a few centuries the Black Sea and the Western Mediterranean became familiar and profitable regions, enlarging the resources available to civilized man.

Every new frontier presents the same problem of vision and risk. According to Herodotus, the Pharaoh Necho of Egypt sent a crew sailing around the continent of Africa about 600 B.C. If Egypt had followed through with further exploration and economic development, Africa today might be a highly developed continent. But the vision was lacking, and no civilized man again saw South Africa for 2,000 years. Leif Ericson discovered America 500 years before Columbus, but apparently the Vikings did not have the vision to see anything worthwhile on that vast, empty continent, and so history waited for another half millennium.

Even our own Western frontier seemed so unpromising just 157 years ago that the Emperor Napoleon, hard pressed for money, sold all the land from the Mississippi River to the Continental Divide to the young United States for only 23 million dollars, including interest. With that one purchase of the unknown Louisiana Territory, President Jefferson doubled the size of the United States and changed the course of history. Thomas Jefferson did not know what he bought, but he sent Lewis and Clark to find out, and private traders, homesteaders, and businessmen followed in their path to turn the empty wilderness into the heartland of a great nation.

Even in our own time, we have had prominent men who stated that airplanes would never fly faster than sound, that intercontinental missiles could not be developed, and that space flight is nothing more than a comic-strip fantasy.

I am reciting this ancient and recent history to be sure that we have the right perspective. When a new frontier is opened, the new territory always looks vast, empty, hostile, and unrewarding. It is always dangerous to go there, and almost impossible to live there in loneliness and peril. The technological capacities of the time are always taxed to the utmost in dealing with the new environment. The explorations require huge investments of both public and private funds, and the returns are always hazardous at first. A few enterprises succeed fabulously, but many fail through inadequate planning or bad timing. The organization, capital, and equipment required for the first exploratory efforts are so large that people tend at first to think only in terms of governmental and military action; and only later do they conceive the new territory as simply an extension of their present territory and their present economy.

It takes an immense effort of imagination for the citizens to see beyond these initial difficulties of opening a new frontier. No one would pretend to foresee all the economic, political, social, and cultural changes that will follow in the wake of the first exploratory shots into space, any more than the people in the days of Columbus could foresee the twentieth-century world. But such an effort at prophetic imagination is what is required of us as citizens, so that we will not, like Leif Ericson, leave the making of the future to others.

The most important long-term impact of the new space capabilities, therefore, is that they open up a new frontier for exploration and economic development. From the businessman's viewpoint, this spells risk and opportunity. But there will be other effects on the nation's business life.

Accelerating Technical Progress

The effort to explore space, and to keep from falling behind other nations in this area, will accelerate the progress of many present-day industrial technologies. The discoveries made in the race for space will have many applications outside the space program. At the same time we must remember that the space technologies are not being made out of whole cloth; they are for the most part extensions of industrial technologies developed for commercial and military purposes.

Electronics, for example, is a technology that has been under constant development and use for four or five decades; but the needs of the space program have put extra pressure on the drive for miniaturization. This will be valuable in such important commercial fields as communications, data processing, automation, and even home appliances.

The space program will also help to accelerate the development of new or unusual power sources, such as the fuel cell, thermionic converters, the MHD (magnetohydrodynamic) process, and vastly improved nuclear power sources. It will undoubtedly help to overcome some of the limitations in materials that are now hindering progress in many industries, by forcing the development of high-temperature, high-strength materials.

The work on rocket engines is bringing in new knowledge about chemistry and combustion. Cryogenics, the study of what happens to matter and energy under conditions of extreme cold, will unquestionably move more rapidly because of the space program. Medical and biological research are being accelerated by the need to know how to sustain life in the airless, weightless, cold, and hostile conditions of space. This listing of the various technologies that will benefit from the space effort could go on for a long time. But the examples I have cited show that many early and practical gains will be made by putting to work in industry the knowledge gained in space-oriented research.

In Philadelphia, General Electric conducts two different businesses. One of them is the development and production of switchgear for the electric utilities and industrial customers. The other is the Missile and Space Vehicle Department. The two would seem to be unrelated, but in some instances they are mutually supporting operations. In the Switchgear Laboratory we have costly special facilities for producing very large electric arcs. The Missile and Space Vehicle Department arranged to use this same equipment to generate the plasma conditions that are necessary to test nose cones and other re-entry vehicles. In the course of their work on re-entry vehicles, the scientists of the Missile and Space Vehicle Department have come up with some rather unorthodox approaches to the control and use of arc phenomena—which are at the very heart of electric power control technology. Thus space research and industrial research mutually support each other.

Impact on Business Thinking

Another beneficial impact of the space adventure will be its impact on the habits of thought in business and elsewhere. It will encourage bold, systematic planning of a long-range nature by forcing men to envision the means of coping with a completely new and hostile environment. It will likewise force long-range speculation and planning in the political, economic, military, technological, and social areas to anticipate future applications and effects of space technologies. Already we are seeing the growth of novel planning organizations made up of physical and social scientists, engineers, financial experts, and professional managers. Their whole output consists of ideas, forecasts, and evaluations of future possibilities. Such planning organizations are already in being in other companies besides General Electric.

The space frontier will inevitably increase the scale of thinking and risk-taking by business. When we are dealing with space, we are dealing with a technology that requires a planetary scale to stage it; decades of time to develop it; and much bigger investments to get across the threshold of economic return than is customary in business today. Business must now think in international terms, and in terms of the next business generation. It must step up to the big risks with the same vision that enabled an earlier generation of builders to push railroad tracks out across the wilderness and lay the foundations of our modern economy.

Another beneficial impact of the space adventure, already visible, is that it is encouraging the creation of more businesses of all sizes. To those who insist that small business is declining, I only ask that you leaf through the advertisements in the technical magazines to see how many thousands of new businesses have sprung up to handle the requirements of modern technology. Most of them are suppliers of specialty equipment for the larger concerns that have responsibility for major components and systems. For, in addition to more small businesses, the space adventure also creates the need for much larger enterprises than are in existence today, with financial and technical resources commensurate with the tasks ahead.

Unquestionably, this need for much bigger companies and groups of companies will run head on into the prejudice of those who cherish an irrational suspicion of any business larger than a family grocery store. But such prejudice will not survive very long in the Space Age. The space challenge will, in fact, force government and business to reassess their relationships, in view of their inevitable partnership in a lengthy enterprise. This is good, because it will force us to face the need for new relationships in a new world.

Thus far I have dwelt on the effects that will be primarily beneficial. But the venture into space has its negative side, too, and all of us as citizens should be aware of it.

A Danger: Government Domination of Industry

It appears that the exploration of space is going to depend, for many years, primarily on government financing and hence government direction and control. That will be true because the exploration of space offers relatively little commercial opportunity for private business in the years immediately ahead.

If the space effort were only a minor activity, the dependence on government financing and control would have less economic impact. But the fact is that the military and peaceful needs of the space program are already employing a significant percentage of the industrial work force, and will make up an even larger proportion of total employment and production of the country as the years go by. The aircraft industry, for example, is broadening its scope to include missile and space technologies. Much of the electronics industry is devoted to missile and space needs. The communications, chemical, and metallurgical industries are increasingly involved. These industries are already among the largest employers in the United States, and they are the major employers of the nation's technical manpower. Hence we are not speaking of a minor element in the national economy, but of its leading growth industries. These industries are subject to ever-increasing government influence by way of government contracts. And the Space Age is only beginning.

Research and development are also drifting under government control. For reasons of defense and space exploration, the Federal government has become the nation's primary sponsor of research and development in practically all of the new technologies. The area of independent research and development by private industry and private universities is becoming proportionately smaller. This means that the pace and direction of progress in most of the leading technologies are substantially under the control of government agencies.

The situation is made worse by the fact that, under the National Space Act, any inventions first applied in connection with a Federal space project, regardless of how they were developed, automatically become the exclusive property of the government. This greatly reduces the incentive for the company that made the invention to develop it for the benefit of consumers and industrial customers. It does not take a prophet to see the long-term effect of this situation on the future independence of the universities and the future vitality of private enterprise.

The patent provisions of the National Space Act are based on the unusual patent provisions of the Atomic Energy Act. But it is important to realize the vast difference between the development and production of nuclear material and that involved in the exploration of space. The development of nuclear material is a relatively clearly defined, closely integrated area of technology. On the other hand, space exploration involves the utilization of nearly every field of science and technology known to man. In the case of atomic energy, it has been possible to erect certain reasonable boundaries around the government's research and development activities, while in the area of the so-called space sciences this is totally impossible. It is the confusion of these two types of technologies which has led to the National Aeronautics and Space Administration Patent Clause which is so unworkable and poses such a threat to the independence of private enterprise.

Government Facilities Expanding

An even more disturbing effect of the growth of government-sponsored research and development is the temptation for the Federal government to build its own facilities and personnel in the technical fields, or to establish so-called nonprofit organizations which are totally dependent on government contracts. However generous their motives, these nonprofit organizations are usurping a field traditionally served by private consulting firms and producer companies, and hence are little more than a blind for nationalized industry competing directly with private enterprise—on a subsidized, non-tax-paying basis. Since the space effort will, for a long time, be primarily a research and development effort, this tendency could lead to an unexpected, and perhaps undesirable, build-up of government-controlled facilities. Looking to the future, when the space frontier has been explored and is ready for economic development, we might well find the area pre-empted by the government, which would then have most of the personnel and facilities available. This would leave the nation almost no choice except to settle for nationalized industry in space.

Please understand me. I am not saying that the people who work in government arsenals and laboratories or nonprofit organizations are any less competent or dedicated than those who work in private industry. I am not saying that government laboratories should be dismantled, because many of them serve a useful function. The government needs a certain number of experienced technical men to help make realistic choices as to future missions, to set high standards of performance, and to provide technically sound policy guidance. That cannot be done by men who are not actively engaged in space research and development. Hence a certain percentage—perhaps as much as 5 per cent—of the technical work of the space program is best done in government laboratories. They should be well staffed and well financed.

But we must recognize that there are growth tendencies in these government agencies that could overexpand under the pressures of the space program, unless proper safeguards are established. As we step up our activities on the space frontier, many companies, universities, and individual citizens will become increasingly dependent on the political whims and necessities of the Federal government. And if that drift continues without check, the United States may find itself becoming the very kind of society that it is struggling against—a regimented society whose people and institutions are dominated by a central government.

To some of you, and perhaps most of you, this may appear to be an alarmist position. In my preparations for writing this chapter, I was surprised to learn how many people have simply taken for granted that space is a natural domain of the government. Almost no one had looked forward to the time when space has been explored, to ask whether its development shall be under our traditional competitive enterprise system, or whether—unlike the rest of the economy—space shall be preempted for government enterprise. And practically no one saw the relationship of these assumptions to the world struggle between free societies and those that are government-controlled.

Principles of a Free Society

It is important, as we undertake this long and expensive race into space, that we as citizens bear in mind the kind of society we wish to maintain and develop. As I stated at the beginning of this chapter, we in the United States have a political and economic system that, in spite of its imperfections, is the most productive and admired system in the world. It not only preserves human freedom, but utilizes it as the key to social and economic progress. We want to maintain this free society in a world where many factors are pushing the nations toward statist systems.

One of the basic principles of the American system is the principle of decentralized power. This country has what is known as a "pluralistic society." There is not just one basic source of initiative and decision here, as in the government-controlled societies. Rather, in the United States, there are many competing points of initiative, risk, and decision— and that is the secret of this nation's drive and creativity.

There are millions of private institutions—colleges, churches, business firms, and associations of all kinds—who do not report to a common center of decision, but operate independently and competitively, trying to excel in their own limited areas of interest. In the government area, there are many governmental units at state and local levels that do not report in a chain of command up to the Federal executive, but rather have their own limited areas of authority. Even our national government is a government of limited constitutional powers, which, for further protection of the individual, is divided into separate executive, legislative, and judicial branches.

And in our society, final authority for nearly all decisions rests with the individual citizens. Through the political system, citizens elect and remove their government representatives. Through the market system, individuals make personal economic decisions as to what to buy, where to work, how much to save, and where to invest. By this intricate system of decentralized personal decisions, the people direct—with a precision that no centralized system could ever hope to match—the course of the economy, the allocation of resources, and the character of national life in accordance with their own personal wishes. The people, not some self-selected government elite, decide the course of the nation. This decentralization of power is the strength of the free society, and it must be preserved.

The Competitive System

In the economic sphere, the principle of decentralized power expresses itself in the system of competitive private enterprise, operating in a basically free market. The competitive system offers many advantages that are not available to centrally controlled economies—and these advantages should be utilized in the space effort and the defense program, as well as in the rest of the economy.

The competitive system, with its profit-and-loss disciplines, puts men and companies to the test as no other system does. It rewards the creative and the efficient. It penalizes the unimaginative and the inefficient. It provides an incentive for risk not only on the obvious ideas, but also on the "long shots." It provides a natural and effective system for the elimination of failure, complacency, and delay. At its best, the competitive economy has a vigor, diversity, creativity, and efficiency that no controlled economy can match.

What is the proper role of the government, in relation to the economy? Basically, it should provide an orderly political setting that encourages individual initiative and competitive private enterprise. It should provide the regulation necessary to keep the economic system competitive—as it does through vigorous enforcement of antitrust laws and other trade regulations. Government should do for the citizens, at their expense, only those things that the citizens cannot do for themselves through their private institutions. Thus, governments provide certain community services—locally, wherever this is possible, but federally, if necessary.

National defense is necessarily one of the services provided by government, but in an age when military power depends less on standing armies than on technological maneuver, the role of private industry is vital. The exploration of space is a project that proceeds from a mixture of strategic motives, including military, political, economic, and ideological. It, like the national defense, will necessarily require a close partnership between the Federal government and private industry. Each of these partners must perform its appropriate role, and the problems and delays occur where government tries to do the managerial and technical work for which industry is best qualified, or industry tries to take on the functions of government.

Meeting the Soviet Challenge

The rapid progress of the Soviet Union in missile and space technology demonstrates how a controlled economy can ruthlessly concentrate major resources in a particular field of technology, and achieve rapid results—while neglecting other fields and keeping the population at a low level of living. Some have taken this to mean that the United States, in order to move out ahead of the Soviet Union in space technology, must adopt something like the Soviet method of strict government control of that technology. In my opinion, such an imitative procedure is doomed to fail.

The United States has its own more effective way of concentrating efficient effort on a technical project of importance to the national security. And that is for the people, through government, to determine the objectives to be attained, and then to turn most of the technical work of achieving those objectives over to the private firms that have the managerial and technical capacity to get the work done—using competition and profit-or-loss incentives to the maximum.

It has become the fashion, since the launching of the first sputniks, to exaggerate the Soviet achievements in space technology. Admittedly, the United States got off to a late start, particularly in the drive for high-thrust rockets. But once we were aroused to the potentialities of space, and recovered from the panic that produced the first series of propaganda failures, the true resources of the United States began to tell. The acceleration of the Atlas and Polaris programs astonished even ourselves, and showed that short lead times are not a Russian monopoly. The United States—even without high-thrust rockets—already has more scientific satellites in orbit, sending back more useful information, than the Soviet Union. When the national need is clear, the partnership of government and industry in the United States can work technical miracles.

But we have no reason to feel complacent. Our present system of defense and space activities still suffers from half-hearted incentives and excessive bureaucracy. This country can surpass the Soviet Union in any technology it selects—if it will use, rather than suppress, its basic strength.

Therefore, it is my view that national economic and military progress will be faster and more solid, and the freedoms we cherish will be preserved, if competitive private enterprise does just as much of the nation's scientific and technical work as possible—and government provides the legal and policy framework to stimulate outstanding technical performance.

Three Stages of Development on the Space Frontier

On the basis of these principles, let us now attempt to foresee the general outlines of the venture into space, and try to determine the specific roles of the government and private enterprise.

The exploration and use of space, like any other exploitation of a new frontier, will probably proceed in three main stages (Figure 49): the stage of exploration; the stage of economic development; and the stage of mature economic operation.

I do not mean that these will be entirely separate periods, or that after, say, 40 years, exploration will end and economic development will begin. Rather, it will be an expanding picture.

First the area of space near the earth will be explored (Figure 50) until it becomes sufficiently familiar, and apparatus becomes sufficiently reliable, for such first industries as satellite communication to be established. We are already well into this phase, and several companies— including General Electric—are already exploring the feasibility of establishing a commercial satellite system for long-distance communications.

Exploration does not stop, but moves outward to the inner solar system (Figure 51), which includes the moon and the nearby planets Mars, Venus, and Mercury. Economic development follows in the wake of exploration.

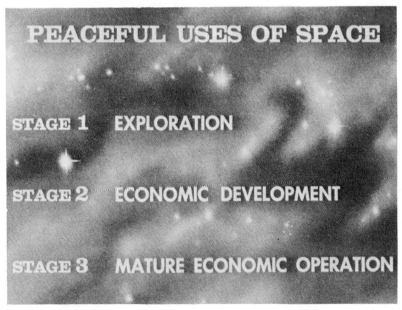

FIGURE 49. Three stages of development on the space frontier.

Before long there may be three phases operating simultaneously. The major area of exploration by that time (Figure 52) will probably be in the outer solar system, including the less dense planets Jupiter, Saturn, Uranus, Neptune, and Pluto. Preliminary economic development may be starting in the area of the moon and the nearby planets. And space immediately around the earth—where most of the present satellites are in orbit—will be entering into mature, systematic economic operation with such commercial industries as long-distance communication on a large scale, private weather forecasting and modification, and terrestrial rocket transport.

Generally speaking, the exploratory stage is likely to be government-directed, with substantial industry participation; the stage of economic development will be marked by government phasing out and commercial industry phasing in; and the stage of mature economic operation will—if private enterprise is to survive in the space industries—be primarily based on private ownership and operation under suitable government regulation, including some form of international law or agreements.

Now let us look at each of these three stages in some detail.

FIGURE 50. At first, exploration must take place near the earth.

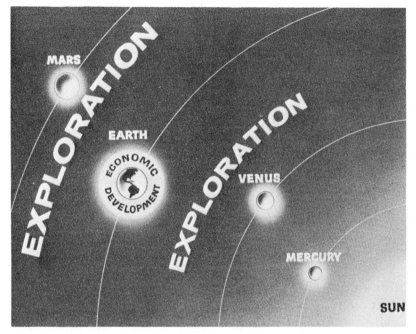

Figure 51. Exploration moves outward to the inner solar system, and economic development begins in the space near the earth.

Stage 1: Exploration

The first era, already launched, is the exploratory period of sending out satellites and other space vehicles, at first unmanned, then manned, to see what is in space. Basically this is the stage of scientific research, bringing back as much scientific data as possible. It will not offer many opportunities for commercial (as opposed to governmental) business for private firms.

The scale and character of this exploratory space activity is indicated in the 10-year program recently submitted to the Congress by the National Aeronautics and Space Administration. This program envisions the launching of 263 exploratory space vehicles in the next decade, at a cost of about 15 billion dollars. That is the estimated launching cost, and it is not clear whether the estimate includes expenditures for constructing and equipping laboratories and bases, for manning such facilities, and for research and development costs of newer space vehicles

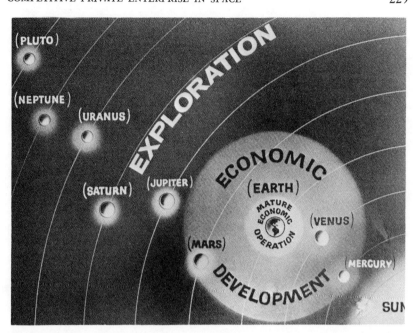

FIGURE 52. Major area of exploration then moves to outer solar system, economic development begins in the inner solar system, and the space near the earth enters into mature, systematic economic operation, with such commercial industries as long-distance communication, weather forecasting and modification, and terrestrial rocket transport.

and propulsion systems in preparation for the following decade. It appears that the annual budget of the National Aeronautics and Space Administration, approaching 1 billion dollars in fiscal 1961, could in a very few years run over 3 billion dollars a year if its program is approved.

The scale of this exploratory work, the cost of it, the lack of any financial return for a long time, and the extra expense of hurrying because of international power politics, almost necessarily makes the exploration of space primarily a government-sponsored and government-financed operation. It is useful to remember that the voyages of exploration in the fifteenth and sixteenth centuries, opening up the Americas and the Orient to European development, were also government-sponsored. But the successful economic development was done in the follow-through period by private traders and colonists—at first with direct government support and sponsorship, and later with the governments serving only to maintain order and provide military and naval protection. On the space frontier, the scientific voyages of exploration will also be government-sponsored and financed.

However, the management and operation of these exploratory operations should be done primarily through government contract by private firms, with competitive incentives for superior performance and penalties for failure. Private firms and private universities should design and produce most of the apparatus required to get there and do the exploratory work.

This approach will not only utilize the most experienced scientific and technical organizations in the country, but will also accomplish the objective faster and more economically, and will help prepare the companies for the day when commercial businesses can be conducted utilizing space technologies.

Private Space Facilities?

Even in the exploratory phase, must we necessarily assume that all the major facilities should be government-owned? As the years pass by, and space apparatus becomes more reliable, and the work of obtaining scientific data from space acquires a more routine character, certainly many of the necessary operating facilities could be put on a self-liquidating, private-industry basis.

Probably the first opportunities for private investment will come in the commercial use of satellites. I have already indicated the interest of private companies in the development of a satellite communication system.

Looking even farther into the future of space exploration, perhaps there would be economic justification for a privately owned launching service that would put objects into space for the peaceful purposes of friendly governments, international agencies, industry, and the universities. The private company would put so many pounds of payload into such-and-such an orbit, at an agreed price. At present, the idea has little appeal, since the government's Scout program will for a time offer this service free to other nations. But as the number and variety of scientific space launchings increase, and other nations no longer wish to be dependent on the United States government, the possibility may be worth exploring. Perhaps the other nations will not wish to invest in duplicate facilities for what is, after all, an operation of planetary scientific significance. This commercially operated launching service would, of course, be strictly for nonmilitary purposes.

It might be objected that foreign governments would feel uneasy about centering their peaceful space programs around launching facilities owned by an American company, for reasons of national pride and national security. But that question could easily be resolved by having the service owned and operated by an international group of companies—including companies from most of the countries likely to use its services.

The base itself, from which the commercial launching service would operate, might be modeled after a port authority. Such a nonmilitary, international space port could develop as a center for many private enterprises related to space operations. These might include service and maintenance facilities; data-processing services; space communication centers; laboratory facilities; standardized equipment for satellites and other space vehicles; fuel supplies; medical services; biological services; and general supplies.

Moving away from the idea of a commercial space port, must all future tracking stations, observatories, and data-processing stations be government-owned? How about experimental stations for the simulation of space environments? How about laboratories and stations actually constructed in space? Or will privately owned facilities one day offer these services on an international basis to governments, industries, universities, and international agencies?

In a preliminary way, General Electric and other companies are indicating their answers. General Electric is building a new Space Technology Center near Valley Forge, Pennsylvania. It is being built entirely with private funds. These facilities will be expanded as the needs become apparent. For example, the company is studying the feasibility of constructing a space environment simulator, duplicating in one 30-foot sphere many of the conditions of extreme cold, vacuum, and radiation that will be encountered in space. The need for a number of such space simulation facilities is not only predictable, but increasingly urgent. There is no reason why the taxpayers should have to build all of them if private industry is willing to make an investment and the government is willing to use private facilities as the major customer in the years immediately ahead.

The Valley Forge facilities, although they will cost several million dollars, are only a small investment compared with the space facilities and launching services I was suggesting as possibilities earlier. But they are a start in the direction of private investment in the economic development of space.

The Monopoly Question

One of the standard criticisms of the entry of private business into fields requiring such high capital investments, and depending so much in the beginning on government business, is that the government may actually be helping to establish a private monopoly. For example, some years from now there may be enough business for one commercially operated launching service, but not for several competing facilities. In this case, it is argued, you do not get the advantages of competition, so why not establish a government facility in the first place?

There are three good answers to this argument. One is that the private facility, even if it is temporarily a monopoly, still has to meet the discipline of earning a profit and avoiding a loss, and hence would be more efficient than a government facility; two, the government would in effect regulate this monopoly because it would be the major customer for some time; three, the monopoly would only be temporary. It would serve the important function of laying the groundwork for competitive private enterprise in the particular field, and keep it from being preempted by government enterprise.

The story of transoceanic air transportation is an illuminating example. In the early 1930s, there were no transoceanic airlines. But it was obvious, technologically, that scheduled transatlantic and transpacific air traffic would some day be possible. To assure leadership for the United States in this important field, Pan American Airways explored the routes and established scheduled transoceanic service. At that time there was no American flag competition in transoceanic air travel, and government airmail contracts were granted to help make the business possible in the early years. This proved to be a wise policy, because United States leadership in international air transportation is now a fact. And Pan American has long since ceased to be the only transoceanic service. International air travel is now a vigorously competitive business, providing reliable, low-cost service for millions of passengers every year.

Surely this instructive example shows how industry and government can cooperate to assure that United States leadership will be established in the peaceful uses of space, on the competitive private enterprise basis that assures efficient low-cost service for the public.

The challenge, of course, is primarily to industry itself. Private enterprise must have the vision and courage to encompass its emerging opportunities, invest the capital, and work for the legal and policy changes essential for success.

Stage 2: Economic Development

As the exploration of space pushes farther and farther out, the "inner space" near the earth becomes sufficiently familiar for the next stage to begin: the stage of economic development.

This stage is a prelude to the time when space will be as familiar as land and mature commercial businesses can be conducted. The developmental period will be a period of infant industries and expensive risks.

Most likely the first businesses suitable for commercial operation, using space technologies, will be world-wide communication by satellite, private weather forecasting, and high-speed earth transport by rocket.

Businessmen who believe in the competitive private enterprise system are generally opposed to subsidies, with two classical exceptions: defense industries and infant industries of national importance. Most of the early space businesses will qualify on both counts.

Here, in my opinion, are the appropriate roles of government and private enterprise in starting these early commercial businesses in space.

The preponderance of the research and developmental work and special facilities should at first be financed by government because of the national interest in establishing United States leadership. But the companies concerned—in their own interests—should also invest in appropriate facilities and manpower to conduct research and development. As an area of space becomes familiar, government-financed research and development should be shifted outward or to other projects of national interest, and private capital should complete the developmental work.

In these areas with commercial potential, the government should avoid the temptation to build operating facilities (under the guise of demonstration units) that will tend to pre-empt the field for tax-subsidized government enterprise and prevent the establishment of private facilities. For example, if in the 1930s the United States had established a nationalized airline instead of helping Pan American to lay the groundwork for international air travel, it is likely that international air travel would still be a government monopoly as far as the United States is concerned. The public then would not have the advantage of many private airlines competing for their transoceanic business.

Private industry should move as fast as possible to establish these early space businesses, so that the government can shift its efforts to the many other areas of exploratory work.

Stage 3: Mature Economic Operation

Finally, we come to the stage of mature economic operations in space. As areas of space become familiar, and businesses become established, the government should phase itself out of this area of economic and technical work and do all it can to encourage the growth of a vigorous competitive private-enterprise economy utilizing space resources.

What these commercial businesses in space may be, no one really can say because we know so little about the potentials. Usually one of the first commercial operations in a new frontier area is to bring back raw materials not available in the home land. It is entirely possible—although many experts now think otherwise—that new or rare minerals and chemicals will be found on the moon and planets, and among the asteroids, that will be unexpectedly useful here on earth. As space transportation becomes more reliable and less costly, and the means of sustaining life in space are developed, it may become economically feasible to mine and bring back these rare metals and chemicals, just as it was worthwhile to haul incense and silk halfway around the world in slow sailing ships 500 years ago. Some have suggested that, as techniques advance, it might be possible to mount rockets on an asteroid of pure iron or pure nickel—if such exist—and orbit the huge mass of metal down to earth. With high-grade earth sources of metal becoming scarcer with each passing decade, the idea of bringing in a new Mesabi from outer space may be less fantastic in 30 or 40 years than it is now. Perhaps the radiant energy of space can somehow be controlled and made useful. Perhaps new medical resources, or new food resources, can be found on other planets. It does not seem probable, but who is to say? Space travel itself, at first for research and then for commercial purposes, may well develop into an economically profitable competitive business. No one can predict how, or even whether, outer space will offer resources that are useful to man; but some very unpromising frontiers in the past have ultimately been tamed.

As the economic resources of space are developed, we must realize that our private enterprises will be competing with the state-operated systems of other nations. But that competition between socialist and capitalist economies already exists here on earth. By the time a mature economy has evolved in space, wholly new forms of economic organization may have evolved on earth—hopefully in a free rather than a regimented context. International laws and treaties with respect to space will be developed at the same time, and we must see that they are compatible with the efficient operation of competitive private enterprise in space.

Legal and Policy Changes

Now, finally, I would like to outline what changes should be made in the government's legal and policy framework to enable private enterprise to make its maximum contribution in the space venture. I am not, of course, prepared with a comprehensive legislative program for the Space Age. In fact, one important principle is that we must not regard space as a completely different area that will require a major break from our established legal and political traditions. The space program must not be used as an excuse for debasing, distorting, or fundamentally changing the nation's distinctive political and economic institutions; as a means, for example, of breaking down the patent system, or of building up a nationalized industry, or of increasing political control over economic life. Instead, we must recognize that the existing system of business regulations and traditions provides a sound basis for the extension of our economy into space for the benefit of all the citizens, as it has already been extended into the atmosphere by way of airlines, broadcasting, communications, and other industries.

To accelerate that proper extension of our economy into space and also realize the military and ideological advantages of leadership in space, there are some legal changes that should be made.

NASA Patent Provisions

The patent provisions of the National Aeronautics and Space Act of 1958 may prove to be a substantial deterrent to progress in space.

Furthermore, they could weaken and ultimately destroy the long-established patent system which has proved to be a key factor in the growth of the nation's economy.

The present provisions of the Space Act, like the patent provisions of the Atomic Energy Act, depart radically from the patent policies of the Defense Department and of industry generally by requiring that the government receive title to all contractor inventions, subject only to a possible waiver at the discretion of the administrative authorities. From our own experience in General Electric we know that such hasty decisions that violate well-tested practices to meet alleged emergencies or special circumstances usually end up doing more harm than good.

Testimony before Congress indicates that under the policy of uncertain patent protection, companies have been hesitant even to accept contracts for space projects that have interesting commercial possibilities. The witnesses made this point: If a company has invested much of its time and money in developing a certain technology, is it realistic to expect the company to liquidate this investment by turning its knowledge over to its competitors by way of the Space Agency?

As it is, only the most readily adaptable inventions made under Space Agency contracts will be developed for consumer and industrial use. Many other inventions requiring greater pioneering effort will lie undeveloped because the inventor company has no incentive to invest in the development of a commercial market; his competitors could easily put a copy on the market just by getting a free license from the government.

We must ask ourselves: Can this country accept the loss of the historic patent incentive and yet win the space race?

Fortunately, a subcommittee of the House Committee on Science and Astronautics has recognized the problem and has developed what appears to be a sensible legal and administrative solution. The Congress should correct the situation as quickly as possible.

Incentives for Performance

Since so much of the initial work in the exploration of space must be done under government contract, the government should realistically offer economic as well as patriotic incentives for outstanding performance in the public interest. Under the government's customary cost-plus-fixed-fee approach, outstanding performance or economy offers little immediate reward. For highly complicated research and development projects, competitive bidding on a straight price basis does not provide the answer. Too often, a contractor would be tempted to submit a very low price and then devote himself to doing the minimum necessary to fulfill the contract. Instead, contracts must be negotiated that offer exciting incentives for exceeding agreed standards of performance, economy, and speed—and sizable reductions in fees in case of failure.

For example, General Electric's Missile and Space Vehicle Department has a contract with the AMC Ballistic Missiles Center that moves in the right direction. It is a contract to come up with the next generation of nose cones, a 2-year research and development contract. The target fee for performing the work was negotiated with incentive provisions that provide a substantial increase in fee for exceeding certain measured performance standards, time schedules, and cost targets. And its fee can be reduced by the same amount by failure to meet these standards. Thus General Electric has a real incentive for technically bold, high-quality performance.

Contracts like this will gradually separate the efficient from the inefficient among the contractors. This is real competition. Such profit incentives will not replace but supplement the motives of duty that encourage leading companies to accept contracts and invest in their own research and development facilities.

Indemnity Problems

The Congress must also accept the indemnity problems that may be too big for private companies to carry on some future projects. For example, space vehicles propelled by nuclear power are a future possibility. There is an extremely remote possibility that one of these might fall in a city and do widespread damage to life and property. The ensuing liability suits could wipe out or profoundly cripple any private company that was in some way legally liable for such an accident, even though it could do nothing to prevent it. The government will have to develop some system to limit the private liabilities in the case of these highly improbable accidents, or companies will be unwilling to risk their entire existence by taking such contracts and the public would have no real way to recover damages. The Congress has already recognized the indemnity problem in the atomic energy area and in connection with international air travel.

Antitrust Limitations

The antitrust laws are a vital bulwark of the competitive system and must be vigorously enforced. At the same time, the large and expensive projects that will win space leadership for the United States will require much larger companies, or even teams of companies, perhaps on an international scale, to carry them out successfully. The law should encourage, rather than inhibit, such large-scale undertakings. Section 708 of the Defense Production Act of 1950 already allows exemptions from the antitrust laws for certain voluntary agreements and programs among companies in furthering the national defense. This points the way toward recognizing the need for larger financial and technical aggregations in the Space Age.

International Law

Any frontier must be subject to law and order. Earth boundaries obviously mean very little in space, and hence a structure of international law and treaties must be developed. The maintenance of order in space will have to be the function of an international agency—most likely the United Nations.

Regulation and Licensing

Because of the international and defense complications, it will probably be necessary to have a system of regulation and licensing for private uses of space. The duties of the regulatory agency should be clearly defined to be sure that the regulatory powers will be used only to assure public safety and necessary technical standards. Such regulatory powers should not be so broad that they can be used to control and shape industry in accordance with the personal preconceptions of the staff of the agency.

Policy Direction

The need for speed and efficiency in the exploration of space requires more coherent policy direction from the Federal government. The individuals who hold responsibility in the various agencies appear to be doing their best to bring order out of chaos, but their efforts in some areas of the space program seem to be frustrated by a confused and top heavy administrative arrangement.

I have not studied the organization in detail. I would only suggest that the number of reviewing committees be kept to a minimum because they can become devices for evading responsibility and postponing decisions; that responsibility and authority for decisions rest clearly with the men held accountable for results; and that any incentives to build empires among the agencies be removed by relying primarily on private laboratories for technical work.

Congressional Statement of Intent

Finally, to assure that the public and the government agencies involved have no misconceptions of national policy, it would be worthwhile to have a Congressional statement of intent to use competitive private enterprise to the maximum in the management and execution of government technical projects; and to encourage private investment in space-oriented technologies and businesses wherever possible.

Our Children's World

To sum up, then, the world is extending its boundaries out from the planet into space: a tremendous enlargement of the area in which man will find resources for living. To explore and tame the new space frontier will require a great technological effort. The very effort will force many new inventions that will not only be useful to us in space, but can greatly advance industrial productivity and levels of living in the United States and the rest of the world.

Yet the ultimate question that faces the citizens at the threshold of the Space Age is not whether the technical achievements will be made, but how they will affect human life. Will the drive for space push mankind into a steel trap of regimentation, or will it open up new vistas of creativity and freedom? Will the new, larger world of the future, with its boundaries moving out to the other planets and beyond, be a free world or a regimented world?

The answer to this question, the heritage we leave our children, will be determined to a large degree by how the United States—the world's leading industrial nation—goes about the exploration and development of space. If we go at it by the route of regimentation and government enterprise, if we allow the communist powers to establish our course, patterns will be set that will be almost impossible to break. On the other hand, if we use the strength of competitive private enterprise, we will not only advance faster, but will help to assure that the world of our children will be a free world, honoring the dignity and creativity of man.

11

Man in Space—
Its Challenges and Opportunities

BRIGADIER GENERAL DON FLICKINGER

ASSISTANT FOR BIOASTRONAUTICS

HEADQUARTERS AIR RESEARCH AND DEVELOPMENT COMMAND

UNITED STATES AIR FORCE

M.D., Stanford University School of Medicine. He was Medical Officer-of-the-day at Pearl Harbor on December 7, 1941, and later served in the China-Burma-India Theater. He became Director of Research, Internal Medicine Division, School of Aviation Medicine in 1947 and later was Command Surgeon, 8th Air Force. In 1951 he joined Air Research and Development Command as Director of Human Factors, then became Assistant Deputy for Development and Director of·Research. In 1955 he assumed command of the Air Force Office of Scientific Research, and in 1958 was appointed Assistant for Bioastronautics and Surgeon, Headquarters Air Research and Development Command.

ANY DISCUSSION, at this point in time, of manned space flight, its many problems and potential returns, must, I believe, be approached with a strong sense of both humility and caution. I certainly feel a healthy and understandable measure of the former and will hope to exercise a reasonable amount of the latter. For I doubt that any one person can possibly do justice to the subject in all of its broad ramifications and implications for the future of terrestrial mankind, even though he were endowed with a degree of supernatural prescience which allowed an accurate prediction of things to come in this challenging field of endeavor. Certainly, this person is not so blessed and therefore will avoid any long-distance look into the future, particularly in relation to specific objective achievements by time.

Never before in the history of our race has an era of scientific opportunity opened up with such great promise of fulfillment in new knowledge to the biologic and medical scientist as does this one of space exploration. The heavens from which man has long drawn spiritual references and sustenance may now be prepared to yield at least the beginning of a true understanding of how life began; of the nature of man's thinking and reasoning; and of the means and manner by which he comprehends the phenomena and mysteries of the universe that surrounds him.

Exciting as these opportunities are, there remain other equally urgent, compelling, and to some of us, challenging obligations which we, as medical, biologic, and behavioral scientists, must recognize and accept. And in accepting, do so with the full realization that these other obligations cannot with any degree of conscience be disavowed as a rightful imposition of responsibility on our total group of workers, and simply be relegated to one segment of our resources which happens to include those in the government service. I am speaking now, of course, of the knowledge and techniques which must be gained and applied to the immediate problem of providing for the safe survival and return of man, who wishes to personally breach the space frontier. One can (and cer-

242

tainly many do) argue that there is no good reason yet for putting man into space; that there is no valid requirement for expending the vast sums and efforts involved in such a venture, either from the standpoint of scientific returns or military expediency. It is equally unrewarding and capricious, I believe, to oppose the issue on the grounds that we are being panicked or forced by Soviet achievements into a program which is "beyond the present state of the art" in science and technology, and which therefore cannot help but boomerang as an instrument for improving our national prestige.

The significant point at hand, I believe, is that man now has available the means to put himself into space, and further, that he will not rest or be satisfied until he has done so—regardless of the immediate objectives or future implications. We, as biomedical scientists, dedicated to the protection of our fellow man from disease and disaster in his endless quest toward greater fulfillment of life, cannot fail him in this, his greatest adventure.

Our nation, either individually or collectively, is not infallible. We have embarked upon a program of manned space flight which is considered sound and feasible by many individuals of mature and proven scientific and technical judgment—National Aeronautics and Space Administration, military, industry, and academic. Having been graced by the opportunity to know and work with them in this undertaking, I am thoroughly assured that no American, regardless of national or international political pressures, will be forfeited on the altar of technological prematurity. Whether an American astronaut is first, second, or third into space will, I am sure, be purely a function of the establishment of an acceptable margin of his safety by those responsible.

Let us look, then, at where we are today; something of how we got there; and what we expect to find and accomplish in the near future. I say *near* future with emphasis and purpose, since we truly cannot say how rapid our progress will be in manned space flight until we have successfully accomplished this first big step of putting a man into a terrestrial orbit and returning him safely.

The Basic Tools

The pace of military technology has been tremendously accelerated this past decade, and necessarily so, if we were to maintain our national security. Vastly improved air, sea, and ground weapons—in the form of supersonic combat aircraft, nuclear-powered submarines, and a healthy family of missiles—have all contributed to our strong defense posture. And we are fortunate indeed that through this past decade we have had wise and courageous leaders in our Armed Services who supported, against all opposition, the development of these advanced weapon systems. For from this vast reservoir of military technology compounded from the dedicated efforts of workers in government, industry, and universities have come the knowledge and tools to get on with the job.

Since we are dealing with man and a vehicle, mated together, to perform a mission in a strange and hostile environment, we turn to military medicine for the answers to the question: "How can we provide for the safety and functional effectiveness of our first space pilot?" This particular field of applied biology and medicine has long been the ever-expanding challenge faced by the military bioscientists, and I can assure you that it has not been an easy one. The history of biomedical research in the services is replete with examples of individual workers who with their own bodies probed the limits of human tolerances to potentially fatal stresses, so that combat crews could utilize the full potential of their weapons and have a reasonable chance of survival. Nor have their good efforts been in vain, for, despite great increases in the complexity and performance of our military vehicles, the casualty rate of the human operator has steadily decreased. At the time of Sputnik I in October, 1957, this country enjoyed the enviable position of unquestioned leadership in the field of aviation and submarine medicine. Our human protective devices and systems were (and still are) widely and extensively copied and used by armed forces throughout the world. From this vast reservoir of knowledge and techniques, then, come the basic medical tools with which we are to provide the means for man's survival in space.

The Problem

If we are to understand with clarity the many facets of man's survival and effectiveness in an orbital space vehicle, we must lay a little simple groundwork by defining the characteristics of the basic vehicle and how it is to operate. We might start out by stating simply that to achieve orbital flight we must be capable of imparting sufficient speed and guidance to a man-containing vehicle to counteract earth's gravitational force at a given altitude above the earth's surface, thereby providing for its continued traversal of the flight path without further expenditure of energy or propulsive force. From our knowledge of energy requirements and orbital mechanics we can put some numbers into the problem. Our orbital vehicle must attain a speed of 18,000 miles per hour, or 25,000 feet per second, and be guided into a near circular orbital path at between 120 and 150 miles altitude.

Now, since pounds of payload in orbit are directly related to how much basic boost or energy is required, we must know very precisely what our orbital vehicle will weigh. In the case of our "single-seater" orbital spacecraft we must have something on the order of 2,700 to 3,000 pounds, if we are to accomplish the total mission requirement. Of this total, approximately 450 to 500 pounds represents the minimum requirement for the human and his supporting equipment. What type of power plants do we have available which can accelerate that amount of weight to orbital speed? Actually, there are a number of combinations of basic missiles with upper stages of additional boost which could do this job, but, remembering our astronaut and the margin of safety which we must have, we are met head on with that age-old problem of reliability. I say "age-old" because man has been faced with this problem of reliability from the time he fashioned his first crude fire-making kit and war clubs. How often will equipment do exactly what it is supposed to? This is the index of reliability and we are constantly surrounded by evidence of varying degrees of reliability.

Complexity of design, environmental forces, and the human component are all pertinent to the net degree of demonstrated reliability. How often, if you press the right buttons, will your television set operate satisfactorily? If you commute to work each day by train or automobile, over a year's time how often will you fail to arrive at the desired time? These are plain garden-variety reliability problems, and in the case of our space-capsule booster are of primary importance. We need a missile of sufficient thrust which has been fired successfully enough times that its likelihood of failure with a man on top of it is acceptably small. Additionally, we know that using a series of boosters in successive stages of thrust compounds our problem of reliability, so we hold out for a single booster to do the job, and in this case it turns out to be the Atlas intercontinental ballistic missile.

The Ballistic Approach

We are ready now to consider the dynamic aspects of the total space mission and extract the biomedical problem areas of potential hazards which confront us. First, acceleration forces on the manned capsule as it is pushed to orbital speed or ejected in emergencies; next, the preservation of viability in the vacuum of space, and of effective performance in the weightless condition; and, finally, the heat and deceleration loads imposed upon the man and capsule as it destroys its speed by friction in the earth's atmosphere and impacts back on earth.

When you stop to realize that the manned vehicle must go through a speed range of zero to 18,000 miles per hour and then back to zero again, and that man's fastest speed obtained thus far in research rocket aerospace aircraft (X-15) is around 2,200 miles per hour, the magnitude of this step is more fully appreciated. No parts of the problem are easy, but the aspect which presents the greatest difficulty is that concerned with re-entry back to earth.

When we look at various possible design shapes for our orbital vehicles, we are immediately struck by the fact that even small amounts of lift given to the re-entry body will appreciably decrease the deceleration loads imposed upon our space pilot. Why then do we use a ballistic capsule instead of one with some slight degree of lift, or even more broadly, why don't we put a single-seater aircraft, such as the X-15 or F-104, on top of an Atlas and boost it into orbit? The answer is two-fold and a simple one. First, single-seater aircraft or orbital shapes with lift weigh considerably more than we can presently boost to orbital velocities. Second, owing to our extensive study of launch and re-entry mechanics of ballistic nose cone shapes we know more about the characteristics of a ballistic capsule under these wide ranges of dynamic loads than any other structure. Thus the ballistic approach to man's first space flight.

The Human Hazards

The vehicle, mission mechanics, and environment produce the spectrum of biomedical interests to which we now attend. Let us make a list of these problem areas, categorizing them on the basis of what produces them.

A. Vehicle-produced
  1. Dynamic G forces of launch, re-entry, and impact, and emergency escape
  2. Noise, vibration, and tumbling
  3. Heat
  4. Immobilization, confinement, and isolation
B. Space-produced
  1. Vacuum
  2. Ionizing radiations
  3. Weightlessness
  4. Communications requirement

These, admittedly, are not pure categorizations, but nonetheless will serve our purpose of discussion.

The problem of accelerative and decelerative forces has received extensive study and experimentation in the military services, directed toward the development of restraint and ejection devices to protect crew members from or during aircraft disintegration. The tools for these studies are the long and short acceleration tracks, the large human centrifuges, and the vertical drop towers.

A wide range of these force applications is required covering rates of onset, direction, maximum peaks acceptable, and time duration at any increased G level. Human tolerances are expressed in terms of those force characteristics which can be accepted by the human anatomy without suffering irreversible damage. These tolerances vary considerably according to which direction they are applied to the body. Their endpoints are expressed as either dysfunctions of the vital organs, such as the brain, heart, and lungs, or actual physical deformation of body structures with attendant pain and injury.

If we now take the predicted or measured G loads to which our astronaut could be exposed during either routine or emergency operations and plot these with our established human tolerances, we find that they all fall within the limits of acceptance as we have defined it. I stress the phrase "as we have defined it," for herein lies an area which will require extensive study (and expensive equipment) in the immediate future. Tolerance curves based upon endpoints of reversible dysfunction and injury may be justified when applied to the problem of ejecting oneself from a disintegrating aircraft where death is the alternative, but they are not satisfactory in the case of manned space travel. In this latter case we want to capitalize on the functional capabilities of the human component, so we must re-establish our tolerance curves at lower levels commensurate with comfort, alertness, and functional capability.

Human limits for noise, vibration, spinning, and tumbling have been studied with considerable care and ingenuity, again using a variety of techniques and devices. Here, we are perhaps on less firm ground when we attempt to match tolerance points against predicted or measured forces, for several reasons. First, as yet we do not know exactly what oscillations or gyrations could be set up by the capsule during critical parts of the orbital maneuvers, particularly during re-entry. Second, the physiologic effects of vibration and tumbling—whether in a state of constant or changing G forces—are insufficiently known; and for the very good reason that we do not have the test equipment available to study this mixture of dynamic forces. This, again, is something which urgently needs doing if we are to proceed intelligently with future generations of space vehicles, where we shall be facing such added problems as escape velocities and impacts on the moon.

I will move on to the space-produced or intrinsic biomedical problems since the factors of heat loads and physical confinement can best be covered in context with the problem of providing a livable atmosphere and environment in the capsule.

The vacuum of space is, of course, untenable to any known form of terrestrial life, and before we can think of asking our space pilot to function we must provide him with a livable atmosphere inside his space capsule. A normal amount of oxygen must be delivered to his vital organs and tissues, and this sets minimum requirements for both gaseous content and pressure of his surrounding atmosphere. Noxious gases such as exhaled CO_2 and water vapor must be kept within acceptable levels, and this requires a chemical system which will absorb specific amounts of both per unit of time. Temperature must be maintained within comfort limits during the orbital phase and within physiological tolerance limits during the terrific heat loads of re-entry. True, the bulk of heat will be disposed of by the heat or ablation shield on the base of the capsule, but significant increases of temperature are still transmitted to the inner walls and atmosphere of the capsule from which the human must be protected.

The life support system is designed to provide this "livable" environment and does so with considerable margin of reserve for safety. Our spaceman must wear a full pressure suit for emergency use, but if all goes well with the total system he will be able to "enjoy the ride and the view" with his face plate open and the suit unpressurized. There are two separate gaseous oxygen sources plus a third emergency supply intrinsic to the suit which, if all capsule components fail, would enable him to survive through a period of emergency re-entry. For the time of excessive heating during re-entry, two heat exchangers or coolers can be valved into the pressure suit ventilating system to keep the body temperature normal.

Since the total time in orbit for the first few shakedown flights will be something less than 5 hours, the problem of food and water is not a significant one. However, for the follow-on multimanned space vehicles designed to sustain crew viability and effective performance over a continuous period of not hours, but days, or even weeks, the logistic and functional aspects of the total life support system assume a position of fundamental importance. Let us put it this way: Unless and until we can provide a reliable environmental control system capable of maintaining a continuous atmosphere of sea-level configuration so that pressure suit wear is not required, we will be unable to capitalize on the full capabilities of man as an observer, as an operator, as a repairman, or as a decision-maker, whether his objectives be scientific, military, or both.

It is apparent, then, that we need extensive research, development, and testing to produce a reliable system which will provide for the following:

1. A near sea-level atmospheric pressure which will allow the space crewman to function freely without pressure suit encumbrance.

2. A gaseous sensing and valving system which will provide precise control of a two-gas system, such as oxygen, and nitrogen or helium.

3. A regenerative capability of recycling and reconstituting metabolic outputs of the body in a manner which will allow their continued reuse. The present system of carrying expendable and non-reusable supplies of oxygen, water, and purifying agents, very shortly becomes, in terms of required days of operation, completely unfeasible from a weight penalty standpoint.

There are many novel and intriguing ideas extant on this subject of partially or totally regenerative systems, but I must emphasize that the biological regenerative system using algae is not just around the corner and won't be for some years to come. Neither is the so-called electro-chemical analog of a photosynthetic ecological system just over the horizon. Smaller steps will have to be taken first to provide more efficient oxygen production per pound of weight, to recycle metabolic water, and to reconstitute the CO_2 removal mechanism. While we are engaged in perfecting this system, perhaps—and hopefully—there will be developed vastly improved auxiliary power sources which will enable us to do such things as dissociate water and CO_2 for partial fulfillment of metabolic needs. Of all the biomedical aspects of space flight, the regenerative system occupies the position of primary importance for the simple reason that until a satisfactory solution is obtained, no further progress beyond short-term orbital flights can be made.

Next, I want to discuss the hazard of space ambient ionizing radiations. Certainly one of the most remarkable scientific achievements recorded thus far in our exploration of space has been that pioneered by Dr. James Van Allen and his group in discovering, mapping out, and quantifying the contents of the great radiation belts. Each day we are learning more about the physical characteristics and phenomenology of these areas of entrapped ionizing particles, and it is really premature to prophesy what will be the complete effect of these belts on future manned space travel.

The inner belt, beginning at about 500 to 600 miles altitude above the equator, appears to be rather uninhabitable, being composed in part of fairly hard and energetic protons whose ionizing capability would be only partially attenuated or neutralized by sizable burdens of lead shielding. Whereas the inner belt or "hard banana" is relatively stable both as regards its content and geomagnetic position, the outer belt, a tremendous area with much softer and less energetic particles, fluctuates greatly in size, position, and content, according to variations in solar activities. With technologically acceptable burdens of shielding and with development of accurate solar-activity predictions, it is quite likely that we can traverse parts of this area en route to the moon or Mars without exceeding the human tolerance limits for ionizing radiation.

Since the orbital altitude of the first manned vehicle will not exceed 150 to 180 miles at apogee or highest point, these great radiation belts pose no hazard, the capsule being well below the innermost one. There is, however, the question of the so-called heavy cosmic primary particles which come both from the sun and from outside our solar system and constantly bombard the top of the earth's atmosphere, losing the bulk of their energies at 60,000 to 70,000 feet altitude. Their biological significance is incompletely known but is felt to lie in their capability to alter or destroy living tissue as they pass through the body. Fortunately, under conditions of normal solar activity the occurrence of these heavy primaries is an infrequent enough event as to categorize it as an acceptable risk. The marked increase in cosmic ray impingement on the earth's atmosphere which occurred in 1956 would have constituted a biologically significant dose of radiation to anyone traveling above 80,000 feet altitude, particularly at northern latitudes. But such solar events are relatively infrequent on our terrestrial timetables, and even should a major one occur "out of the predicted cycle" while our astronaut was in orbit, we could bring him back to earth before any irreversible damage was done.

Although the ionizing radiation hazard is considered minimal and quite acceptable during short-term, low-altitude orbital flights, it is obvious that much work needs to be done by both the bioscientist and physicist before we can even define the total problem satisfactorily, let alone work out the solutions. Fortunately, much work is being done in this area in close collaboration between scientists in the government and universities, and there is even the possibility that our advancing particle accelerator techniques will enable us to simulate to some degree the energies of the heavy primaries. Such capability would, of course, open the door to very precise studies on the full range of their biological effects.

Weightlessness, which occurs when the velocity of the vehicle balances out the gravitational force, is a completely unnatural phenomenon for man, who evolved and lives under a constant one-G environment. Human reaction to weightlessness can be studied only for very short periods—60 to 90 seconds—in manned aircraft during so-called parabolic flight paths. However, one of the early Air Force studies in space biomedicine was directed toward the observance of physiologic disturbances in animals launched in Aerobee rockets by a team under Dr. James Henry. The 27-mile-altitude flight path provided a near weightless condition over a period of approximately 5 minutes, and no untoward immediate or delayed effects were noted in those animals which they were fortunate to recover. These experiments, begun in 1949 and completed in 1951, have been duplicated many times by the Soviets and ourselves, particularly since the launching of Sputnik I. At the time this was being written, the most recent was the ride of Miss Sam, a rhesus monkey, in a Mercury prototype capsule being tested in one of the escape configurations.

When you analyze all of the data obtained from these animal experiments, you find pretty much the same results as were described by Henry and his associates in 1951. Namely, for the periods of exposure attained there were found no appreciable disturbances of physiologic function either during or after the weightless exposure. Laika, the dog that went up in Sputnik II, was reported to have undergone the weightless state over a period of from 5 to 7 days without evidence of harm. However, there is question as regards the stability of Sputnik II in orbit, and if there did exist even a slow rotational movement of the capsule, Laika would not have been in a true zero G environment.

Our short-term human experiments have been quite productive in terms of usable findings for the problem at hand. Difficulties in eating and drinking are easily mastered with suitable techniques and devices. No physiologic malfunctions occur and psychomotor tasks, after a short learning period, are performed with required speed, accuracy, and dexterity when the subject is properly fixed and positioned in relation to the equipment. The lower extremities seem much less adaptable to zero gravity, and locomotion—even with magnetic shoes—is difficult. Subjectively, sensations vary from apprehension to euphoria, with the latter predominating. Disorientation and vertigo occur spontaneously in a proportion of exposed volunteers, a small percentage of whom continue to show the same reaction despite repeated exposures. Disorientation, vertigo, and nausea can be induced in a large percentage of the seasoned subjects by making them perform active tumbling maneuvers during the exposure. Almost all show an ability to "adapt" to the weightless state both in terms of performance and in maintaining satisfactory spatial orientation if proper fixation and visual clues are provided.

Recently a volunteer at the Aerospace Medical Center spent 7 days balanced out in a water tank in an attempt to study long-term effects of the weightless state on circulation, respiration, and muscle metabolism. All functions dropped to below the control baselines and we found that he needed appreciably less sleep and food during the exposure. Upon termination, however, we found that he needed a period of several days of reacclimatization before he felt normal while standing or walking in the erect posture. It would be an amusing irony indeed if we found that exposure of days or weeks to weightlessness during space voyages required a period of gradual readaptation to the normal gravitational environment of the earth before returning. Or, looking somewhat into the future, to learn that living in a one-G environment is a factor in the development of some of our poorly understood degenerative changes associated with aging. Many generations hence, it will be of interest to note the life span of human lunar colonists who are born and grow up under the influence of a gravitational force only one-sixth that of our own earth.

All of us biomedical people would perhaps feel better about even the relatively short (3 to 5 hours) period of weightlessness, particularly in terms of its effect upon orientation, if the animals which have been exposed to longer periods could talk and tell us how they felt. Facetious as this sounds, it is nevertheless true that no medical instrumentation or technique is available which can tell us about disturbances in the brain or nervous system which may ensue in the weightless state. Only the voice link with our astronaut will make it possible, through his own reporting, to estimate significant alterations in the subject's sensations and perceptions.

There are excellent experiments in the making which will markedly increase our knowledge of animal brain function and behavior, both in the hypergravic and weightless phases of an orbital mission. One particular group at the University of California at Los Angeles Medical School, under the direction of Doctors French and Magoun and captained by Doctor Ross Adey, has developed techniques which provide a stabilized positioning of electrodes implanted in the brain, which enables them to study and correlate the effects of varying dynamic force fields on both behavior and electrical activity of the brain. The eventual application of this knowledge and the techniques to animal satellite studies should increase considerably our scientific returns from such experiments.

I have mentioned briefly some of the psychological aspects of the weightless state which are only a small portion of the total problem of psychological disorders and hazards of space flight. These are usually spoken of in situational terms, such as physical immobility, close confinement, terrestrial and temporal dissociation, and isolation from fellow man in a reduced sensory environment. A great amount of work has been done by psychiatrists and psychologists using many ingenious techniques to simulate these various aspects of space travel. And while it is true that in some experimental situations producing extreme degrees of isolation and sensory deprivation, some subjects have shown rather severe mental disturbances manifested by hallucinations and panic, I do not honestly believe that such will occur in our early space pioneers. The conditions which actually exist in the space capsule are in no way similar to those which produced the abnormal psychic reactions. More closely simulated inner-capsule conditions have, in well-motivated and carefully selected volunteers, resulted in no abnormal behavior and in quite respectable levels of psychomotor performance over considerable periods of time. The pioneer astronaut will be kept quite busy with many tasks to perform, and this, along with a very active interest in what is going on about him should suffice to neutralize any effects of isolation, confinement, and reduced sensory environment. I personally do not look for any psychological problems to arise until we attain the capability to transport groups of people on relatively long space missions, and here the problem of "forced association" may well pose some difficulties. For the immediate future, however, I feel that the critical biomedical hazards of orbital flight will be experienced more by the body than the mind. Though with the propensity of space to provide many unexpected surprises, as Dr. Dryden puts it, "I probably should be prepared to have this particular prediction come back to haunt me."

My last biomedical hazard listed as space-produced was that of the communications requirement. This cannot quite correctly be classified as a hazard per se, but rather should be considered as creating an uncertain or hazardous situation if adequate communications cannot be maintained. Our requirement for communications, then, stems from two sources, closely related. First is a "safety of flight" factor which consists of an urgent need to know what is happening to our spaceman and his protective environment on a continuous basis. Second, we wish to collect as much scientific data as we possibly can during these early experiments.

We need to sense as many of the vital human functions as we can, along with data on the capsule environment and its control system, and both store the information on board and telemeter it to the ground monitoring and control stations. Finally, the voice transmission link between capsule occupant and ground, in the final analysis, may well prove to be our most important communication facility. Certainly it could be the major factor in making a decision to abruptly terminate a mission.

The whole field of medical electronics as applied to the orbiting astronaut constitutes a most difficult problem with which practically no one is happy. You can start right from scratch with the question of what aspects of physiologic behavior are truly meaningful and valid as indices of significant changes. From here, then go to the problem of sensors or pick-ups which will function reliably over a wide range of conditions and yet allow the man to function without discomfort. Next, come the questions of power requirements, amplification without distortion, frequency and band-width restrictions, and finally that of faithful recording of the events for data reduction and analysis. This is assuredly a most fruitful area for future scientific and technological endeavors on a broad and imaginative front, drawing from the knowledge and skills of a wide range of scientists and engineers. Meanwhile, we will get along as best we can with what we have and trust that our predictions will be sufficiently accurate that we will not be faced with an unexpected misfortune that cannot be explained with the data at hand.

The Next Generation of Manned Spacecraft

Assuming that all goes well on these first orbital flights, where can we expect to go from there and how soon? There are many factors involved in attempting to answer this question, and I am not in a position to consider and evaluate the majority of them. For the foreseeable future, space exploration costs will remain inordinately high, and pounds of payload in space will figuratively and literally be "worth their weight" in gold or platinum. How much of the national income and effort should go into space exploration as compared to all the other government programs requiring increasing amounts of money? What is the relationship between our progress in space science and technology and our national security—our military strength? These are all vital and important considerations bearing on the problem which merit most searching analyses and evaluation at the highest levels of government.

With the assumption that the space program will be nourished and supported as now planned, what can we expect to accomplish when payloads of 5, 10, 50, and 100 tons become available? Here again, I believe, we are faced with the problem of apportioning our space resources to those objectives which appear to offer the most return to our country. Though many deny it, military expediency and the temper of the times might require an all-out effort to place a sizable manned space patrol force into strategically important orbital tracks, or even to go all-out for a manned lunar base. On the other hand, under less international tension, it could well be decided to concentrate on unmanned space weather stations and communication links, and to utilize the payload capabilities set aside for scientific advancement for such intriguing experiments as searching for evidence of extraterrestrial life or its precursors on Mars or Venus. The field of astrobiology, or exobiology as Dr. Joshua Lederberg has named it, contains a great array of exciting possibilities for the major advancement of basic biologic knowledge through the execution of properly designed space experiments. One could certainly not deny the dedicated basic scientists a reasonable and fair share of our space opportunities and resources.

Where does that leave us, then, with the continuance of our manned space vehicle program? I personally believe that it will continue in a healthy fashion with the rate of progress being intelligently paced by our accumulation of new knowledge and techniques. I can easily visualize the next step of putting three or four men in a tandem-type orbital vehicle, with the "trailer" providing work and storage space, and the "tractor" handling the control and energy management functions and being finally used as the re-entry vehicle when the mission is completed. I say I can visualize this next step because the advancements needed in biomedical knowledge and capabilities to support this step are attainable if our total resources—facilities and people—are properly utilized and supported toward the achievement of these objectives.

I see such a multimanned orbiting vehicle as accomplishing some extremely worthwhile objectives, to wit:

1. Provide a testing ground for determining accurately the operational capabilities and maintenance requirements of advanced components

2. Provide an opportunity for an exhaustive and detailed study of man's reactions to the space environment and of his full range of performance capabilities in a space vehicle

3. Provide a base from which significant human observations can be

made and recorded on a wide range of terrestrial, solar, and planetary phenomena

There are many who will argue that all of these objectives can be realized with less expense and more expedition by utilizing ground space-environment simulators and unmanned orbital laboratories, but I will not join this battle because to me it is wasteful of both time and energy. The proof of the pudding is still in the eating, and only time will yield the correct answer. I do not argue against simulators or un-manned satellites in the slightest degree. All that I am attempting to do is to use yardsticks of past experience to estimate the problems of the future, and the time and resources required for their solution. The fact that we are dealing with a manned space craft instead of a manned high-performance aircraft should not require a completely different set of ground rules for carrying out the research, development, and test program required.

Air Force experience supports abundantly two fundamental tenets which underlie the successful development of a progressive family of manned aircraft: first, that basic and applied research in all areas includ-ing the biomedical must be supported on a continuing basis on a time scale well in advance of the vehicular specifications; second, that no amount of preflight simulated testing can prove out the capabilities or deficiencies of the final operational vehicle. It might well turn out that manned space test beds would provide the most economical and reliable means of proving out the worthiness of advanced components and as-semblies. Certainly, as a training ground for space crew members, nothing could ever equal the space laboratory as the ultimate and final proving ground for human space adaptability.

Summary

We are in the barest infancy of the space age, yet even now there are unfolded before us greater challenges and opportunities for mankind than ever before existed. Space exploration will continue and flourish for as long as mankind exists on this earth, and its returns in all fields of science and technology will be practically unlimited. For the medical and biological scientist, the opportunities for gathering new knowledge and making major contributions to mankind are in three categories: first, the biomedical support of manned space vehicles and platforms;

second, the use of space, with its unique physical characteristics, as a laboratory to study fundamental life processes which may well provide the answers to disease processes which have long eluded us in terrestrial laboratories; third, the examination of space and its planetary and galactic interfaces for extraterrestrial life or prelife forms—all the way from a primordial "cell" or substance linking inorganic materials to organic self-replicating life systems, to complex living organisms like ourselves inhabiting other planets. Each area contains its own exciting possibilities and has a proper claim upon our national space resources if the biosciences are to advance on a parity with the physical sciences.

For those of us in aviation medicine, long interested in optimizing the safety and effectiveness of crews operating high-performance aircraft, the challenge of manned space travel is great, and a natural extension of our past interests and work. There can be no doubt that man will develop a progressive family of increasingly larger and more sophisticated space craft just as occurred with aircraft, and it is fairly unlikely that he will invent an automatic robot to replace himself, thereby being denied his age-old prerogative of original discovery in a new frontier.

The question of military or peaceful man in space is not one that need concern us here, for the biomedical problems are the same in either case. Only time can tell how mankind will use this new medium of travel—for his future enhancement or for his destruction. What is important is that our country remain strong and vital on all fronts as we proceed deeper into this exciting new era of the space age. To do this, we must first of all have a national awareness of the magnitude of the challenge with all facts and implications available for honest appraisal. Next, and of even greater importance, is the acceptance of this challenge with the highest order of motivation and courage, reflecting the true spirit of our pioneering forebears, to keep America foremost in space technology and achievements. Finally, we must marshal, nourish, and replenish all of our scientific and technological resources wherever they be, in the military services, in industry, or in the ivory towers of academic life. There is more than enough for all to do, and for the first time in many generations let us hope that we can rise above prejudices and provincial interests and strive as a nation to preserve our way of life for our future generations.

12

Outer Space Travel— What Is and Is Not Possible?

EDWARD TELLER

PROFESSOR OF PHYSICS

UNIVERSITY OF CALIFORNIA

Attended Karlsruhe Technical Institute, Germany, 1926–1928, University of Munich, 1928–1929; Ph.D., Leipzig, 1930. Held research and teaching positions at European universities 1929–1935. Professor of Physics, George Washington University, 1935–1941; Professor of Physics, Columbia University, 1941–1942; Physicist, Manhattan Engineer District, 1942–1946; University of Chicago, 1942–1943; Los Alamos Scientific Laboratory, 1943–1946; Professor of Physics, University of Chicago, 1946–1952; Los Alamos (on leave, University of Chicago), 1949–1952; Consultant, Livermore Branch, University of California Radiation Laboratory, 1952–1953; Professor of Physics, University of California, Berkeley, 1953 to present; Associate Director, Radiation Laboratory, University of California, Berkeley, 1954 to present; Director, Radiation Laboratory, University of California, Livermore, 1958 to 1960. Honors: Doctor of Science, Yale University, 1954; Doctor of Science, University of Alaska, 1959; Doctor of Science, Fordham University, 1960. Awards: Joseph Priestley Memorial Award, 1957; Albert Einstein Award, General Donovan Memorial Award. Member of National Academy of Sciences; American Nuclear Society Fellow, American Physical Society; American Academy of Arts and Sciences; Scientific Advisory Board, Air Force; American Ordnance Association. In recent years he has attracted attention for his role in the practical application of thermonuclear principles in the development of thermonuclear weapons.

THERE ARE two ancient visions—two ideas related to the physical world—which have influenced men for thousands of years. One of them is the ocean. The other is the heavens. Both of them have stood for the concept of infinity.

A few hundred years ago, navigators embarked on the exploration of one of these seemingly unlimited realms. They finally succeeded in circling the globe. We now know that we are living in a very limited world indeed, which scientific and technical progress makes smaller every decade. We are leading a more and more crowded existence— with more and more clear demands that we find ways to live with our neighbors if we are to live in peace on this small whirling globe.

The other of these symbols of infinity—the heavens, or space—is something which is still infinite as far as we know. We know a lot about space but we have barely started to penetrate it. At the threshold of such a new adventure, people naturally ask what is the value of doing this. What will the space explorer bring back? But the experience— the history—of the great adventure and the fateful travel of Columbus is perhaps, and should be, a lesson on how to ask such a question (or not to ask it) and what kind of an answer to expect. When Columbus started on his exploration, he thought he knew what he was after. Namely, trade with China (this, I am sorry to say, we have not accomplished to the present day). But instead, he found a couple of continents—in part of one of which we Americans now live. He found a new world, and the results of his labors were magnificent, and quite different, and certainly bigger and more important than he or anybody else expected.

I have no doubt that the same will be the case in connection with space exploration. With this expectation, I am going to try to sift fact from fancy. I am going to try to make some predictions. I know that in this field it is dangerous to be a prophet, because the development is so very fast that my prophecies will prove to be wrong even in my lifetime. Yet, if I err—and I surely will—I shouldn't be surprised if it would be because my imagination—the thing that differentiates me from all the animals—has been too weak.

What are we to find in this exploration? There is no doubt in my mind that in the next few years, surely before the end of the century, we shall have explored the planetary system. What will be the value of this? Shall we go and live or will our children go and live on the moon, or Mars, or Venus, or the big spaces of Jupiter? I don't believe so. These places from all we know are rather uncomfortable. Of course, there will be energy available; we can carry nuclear energy along, and we can heat or cool with the help of some appropriate apparatus. But let us take the most agreeable of these planets—Mars. Its surface area is not much more than a quarter of that of the earth; it is a little on the coldish side; it does not have much gravitational pull, which might be agreeable; there is a little oxygen, probably not enough for comfort, but useful; there is some water; and there is the suspicion of some life. With all of this, I believe that existence on Mars will be considered a little off-beat as compared to living in Antarctica. So this is not precisely the advantage I am looking for, though I may well be proved wrong.

Again, you might be thinking about going out there and bringing back some valuable things like gold, or uranium, or diamonds. Well, diamonds we're learning how to make down here. As for gold or uranium, it would be much too expensive to transport them back, even if we should ever find them in the pure state. Yet I believe there is something we can bring back—something of which we have far too little and of which there is lots to be found in space. I have in mind a commodity which is pre-eminently adapted to space transportation because it has no weight. It is knowledge. The principal thing we will want to bring back from space exploration is more knowledge.

As one example, consider the tremendous advantage that astronomy will acquire by getting outside our atmosphere. We shall be able to see the universe in all wavelengths—in Technicolor as it were. As a little preview, we can recollect that before World War II, our atmosphere appeared to prevent us from seeing the universe in any wavelengths other than the one little octave between blue and red, which we call visible light. Then, after World War II, another window was opened up: short wave, radar, radiation. This has led to most remarkable discoveries. We have found that the storms on the sun emit such radar waves. We also have found that certain portions of the heavens emit radar. For instance, more than 900 years ago, at a certain distant place in our galaxy—in our family of a hundred billion stars—one of these hundred billion stars exploded and the explosion was duly recorded by two Chinese Royal Astronomers. At the place of this supernova explosion, we now see a turbulent gas mass, and connected with this mass there are magnetic fields which are moving and which interact with the ionized gases that were ejected from the supernova. All this disturbance seems to send to us quite a bit of radar noise, from which we can learn something about the behavior of this unquiet gas mass.

I have already told you that our sun is one star among a hundred billion brethren. They form an enormous system whose longest dimension light would take a hundred thousand years to cross. All of this is arranged in a huge spiral structure. We in our solar system lead in one of the distant branches a rather isolated and suburban existence. But our galaxy is not the only one that is around. Approximately two million light-years from us there is a similar galaxy, the Andromeda Nebula, and then as we go farther and farther, there are millions and millions of others. At a distance of much more than a hundred million light-years from us, two such galaxies happen to be colliding. This collision has produced so much radar noise that from these two galaxies—even though they are so far away—we derive one of the strongest radar sources in all the heavens. There are probably such colliding galaxies even farther than the most powerful telescope can now see. Perhaps farther than the most powerful telescopes ever will see. And thus, by observing these radar phenomena—these phenomena on a new wavelength—we have a new method of peering into the ultimate depths of the universe and perhaps of finding out whether heaven, like the earth, will in the end prove finite, or whether our universe is truly infinite.

These are little pieces of knowledge which we have already picked up. But now by going out of our atmosphere, instead of just seeing a few wavelengths in the visible and a few wavelengths in radar, what else will we see in all the other wavelengths? We don't know for sure, but the prospects are obviously exciting.

By going up to the moon and to the various planets, we shall be able to make new observations of these heavenly bodies. These will not be astronomical observations in the strictest sense. Let me give one brief example. There is a certainty that we are going to land on the moon.

By we, I mean humans. I wish I could say with full confidence that the Americans will be first. Anyway, somebody will land on the moon. And when he lands, one of the first things he'll do, I believe, is look on the moon for earthquakes—I mean moonquakes. And furthermore, if my respected colleague Harold Urey is right, the moonquakes, if any, will not have the same cause as do earthquakes. We have earthquakes down here because the earth is an inhomogeneous structure consisting of a lighter mantle and a heavy, iron core. The light and dense materials are not yet completely separated, and during their sorting out, disturbances occur which are responsible, Professor Urey thinks, for the folding of mountains and for our earthquakes. The moon has a density the same as the mantle of the earth. The same process that we observe on the earth—or rather that we hypothesize on the earth—probably does not go on, on the moon. Also, on that part of the moon's surface with which we are well acquainted, we do not see any folded mountains. We see craters, which may well be due to splashes caused by big meteorites that at one time or other have fallen into the moon. Such craters would remain there like the relatively recent meteor crater in Arizona, not being leveled out by the nonexistent atmosphere and rain on the moon. But in all those ancient craters standing there, we don't see mountain ranges. So perhaps the mechanism which makes mountain ranges and earthquakes does not operate on the moon. It certainly will be interesting to find out.

We may go on to Mars. Mars doesn't have any folded mountains either. It, also, may be in that class of heavenly bodies which are too homogeneous for that kind of violent thrust. We may go to Venus—and Venus is even more interesting. It has so opaque an atmosphere that we have never seen its surface. What is underneath is a secret. We may go farther—to Mercury, half of which is very hot and half of which is very cold. We may go on to the outer planets. I certainly would be interested in knowing what has caused in the last century that huge red spot which appeared one day on Jupiter and then faded, but is still visible in outline. We don't know what it is. We could learn a lot about the structure of these planets. We could learn a lot about many other things.

Let me mention just one additional little idea before we get on with our main subject. There is in space a commodity which we are spending many millions of dollars to get in an imperfect way on earth; namely, vacuum. For many scientific experiments we need a good vacuum, and where could we find a better one than in space. Perhaps certain scientific experiments should be conducted in a space station or on the moon because we have in these places such an excellent vacuum. Perhaps all the high-energy physics will become within a few decades lunar physics, because the proper surroundings could be obtained more easily on the moon. I doubt that will occur, but stranger things have happened.

However, all of this is relatively unimportant. We, as living beings, are obviously more interested in one topic than in anything else; namely, is there any life on the moon or the planets? What shall we find? Well, I certainly don't know. Between you and me I do not believe that the Martians are 2 feet tall and green. I don't know whether there is life anywhere in the solar system except on earth. But I have a strange worry. Life is something so peculiar that (if you try to think about it in a reasoning manner) you certainly should not expect it to exist. What else can be found in the universe? Another form of life or still something else? The last thing that our imagination could have predicted is—ourselves.

Well, the moon seems dead—probably it is. Mars has some changes of color which are suspicious. Also, looking at Mars with a spectroscope one finds strange and intriguing lines which are characteristic of the carbon-hydrogen bond. A carbon bound to a hydrogen—we find plenty of that on earth, wherever there is life and in some places where there are traces of past life, such as in petroleum, which is probably a remnant of very ancient life. This combination of atoms is present on Mars and may well be due to something other than life, but at least there is a suspicion that it is due to life. And since Mars has a little oxygen, a little water, and a temperature which is not quite absurdly low, it may possibly have life. Now, if we get there (and we will), and if we encounter life, I would hazard a prediction. Life on Mars will be so strange, so unusual in appearance, that the first observer will not recognize it as life at all.

I would like now to point out an old and interesting fact. In the eyes of a chemist all living things on earth are cousins. They all are built from exceedingly similar chemical building blocks. Whether it be man, chimpanzee, fish—whether it be a tree, a bacterium, or even a virus— it always contains proteins, and these proteins are made up of chains of different kinds of amino acids linked together. I want to tell you what

an amino acid is and what a protein is, in order to show
how specific this similarity is between all living things.
Figure 53 shows a generalized diagram of amino acids.
C stands for carbon, O for oxygen, H for hydrogen, N
for nitrogen; and R is a radical, or side group, which
is different for each kind of amino acid. In the case of
the simplest amino acid, R is simply a hydrogen atom,
but in most of the amino acids it is some sort of a chain or other
structure consisting of carbon atoms and other atoms. There are some
twenty-odd different radicals that can hang there, and all living things
have these various kinds of amino acids, with these radicals, in their
bodies. A protein is built up out of amino acids by the fabrication of
polypeptide chains of tremendous length, and Figure 54 shows how the
chain bonds are formed. The hydroxyl group (OH) of the first amino
acid combines with one of the hydrogens of the amino group (NH_2)
of the second acid to form a molecule of water. This leaves the carbon
of the first acid free to tie on to the nitrogen of the second acid, and
in this way one of these amino acids gets tied to the next one, the next
one tied to a third one, and so on. It is very likely that the specific
differences between living things are closely connected with the special
sequence in which these amino acids follow each other, whether it be in
a virus or an elephant.

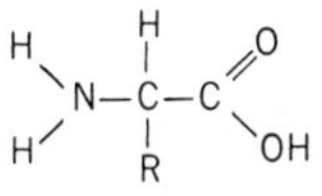

I would like to know whether the Martians are also made up of amino
acids and, if so, whether they are made up of the same kinds of amino
acids in the same way. Let me give one other example to show how far
the similarities go. One very important function of most living things,
as we know them, is to take up oxygen or to give away oxygen. This is
the oxidation-reduction mechanism. This process goes on with the help
of certain pigments—heme in the hemoglobin of our blood, chlorophyll
in the leaves of plants. These two exceedingly important substances

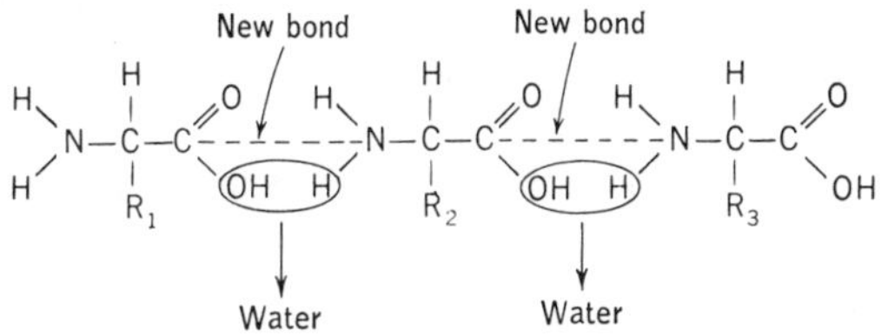

FIGURE 53.
Generalized diagram of amino acids.

FIGURE 54. Formation of polypeptide chain from amino acids.

are vital in the animal and vegetable kingdoms. Now, these two substances have a remarkably similar structure. As shown in Figure 55, the basic unit is a five-membered ring of one nitrogen and four carbon atoms. There are four such rings which are tied together by four carbon bridges. Depending on what the particular compound is, all of the outside atoms have various attachments (not shown) of carbon, hydrogen, and oxygen. And in the middle of the whole structure, there sits a metal; namely, iron in the heme of hemoglobin, copper in sunfish respiratory pigment, and magnesium in the chlorophyll of plants. I would like to know whether this very particular structure—one out of hundreds of thousands of known organic structures—is predominant in the living things of Mars as it is predominant on earth.

If you ask me what life is, my best definition would be: life is a little matter with a great deal of purposeful complication. I would like to know whether the Martian complication is similar to the terrestrial complication. If the answer is yes, we may have a common origin. If it is no, we will have a different origin. And in either case, it will be most interesting. Somehow, I have the feeling that the answer will be more ambiguous and more surprising. But we shall see. Of course, there is the possibility of a real disappointment. Mars may have no life at all. Even so, I am sure that the universe is full of life, is teeming with life. In our own galaxy we have more than a hundred billion stars. Many of them surely will have planets. And some of the planets should be inhabited. And beyond our galaxy there are billions of other galaxies. We certainly are not the only living beings. It would be very strange to believe that. I also doubt that we are the only intelligent beings. But if the universe is indeed as old as it's said to be—almost 10 billion years—and human life spans only the last half-million or million years, there must be others who arrived earlier. And, I would like to know, where are all these other people?

Well, I have given you a partial answer. We are living a suburban, isolated existence and it is quite possible that nobody has yet happened to come by this god-forsaken neck of the woods. To

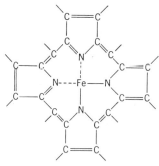

FIGURE 55. Central structure of the heme of hemoglobin.

illustrate this, let me raise the reverse question. I have mentioned that by the end of the century we shall have, hopefully, explored our planetary system, but when shall we get to the stars? The nearest known star is approximately 4 light-years away. Light takes 4 years to go from the sun to Proxima Centauri. According to Mr. Einstein, nobody can go faster than light. Nobody can go even as fast as light, except if he divests himself of all mass. But, we might go almost as fast as light. Four years is a long time, but still it would be quite an adventure to get to Proxima Centauri. However, the technical feasibility of all this is somewhat in question. Let me take the most efficient fuel which has been used today, the kind of fuel which gives the most energy per unit weight. This is the fuel of a fusion reaction. If you will let me make a rocket and postulate that I am very clever so that the machinery weighs practically nothing, and all the weight is in the fuel plus the payload; and if you will let me make the payload about as heavy as the fuel itself, or a little less heavy, then such a rocket could acquire the speed of one-twentieth of the speed of light. A traveler would therefore take 80 years to get to Proxima Centauri, a somewhat discouragingly long time.

I am not saying that we cannot do better—we can. We can use the principles of the multistage rocket. Unfortunately, if you want to go faster than this natural velocity just mentioned, you have to add very great amounts of fuel, and to propel a man and his environment to such great distances will be a gigantic task. I am almost tempted to say that it cannot be done; except that I have had the experience that in most cases where I have been wrong in the past, I have been wrong because I felt that something could not be done. Also, looking back at what people could do 300 years ago, 200 years ago, a mere 100 years ago, I am rather loathe to say what we cannot do 100 years from now. Perhaps it will remain impossible for a long time—perhaps forever—for a person to travel to Proxima Centauri. But I would expect that we can send at least some very light apparatus up there, because we would need proportionately less fuel and the undertaking would not be quite as gigantic. Even that would be wonderful; and at least that apparatus wouldn't get bored, although a great deal of thought will have to go into its construction so that it will remain in good order for many years. In the end, we might get to Proxima Centauri and find that it has no planets. And then, we might have to go farther. And this business of star-hopping might be very, very difficult, tedious, hard work.

Well, do all people in this universe face similarly great difficulties in interstellar traffic? The answer is no. Let us take a look at the metropolitan central area of our galaxy. All galaxies contain a core of stars, where the stars are close together. We can't see the core of our own galaxy because it is hidden from our view by a cloud of dust. If we could see it, it would have a brightness one-tenth that of the full moon. In that core the stars are much closer together. In that core it is quite possible that people from different stars are already merrily exchanging information, perhaps even carrying on some kind of a trade of the sort which I said would be impossible. It is even imaginable—but I hope it's not true—that they will fight interstellar wars. If any of this should be true, we might have difficulties in participating, because this metropolitan region is approximately 30,000 light-years from us, a distance a bit too far for the usual kind of commuting. We might have a better chance of listening in on their radio talk, particularly if we can get out of the atmosphere and listen on all wavelengths. And when we learn what they say to each other, it is conceivable that we shall be grateful for our insular existence. (I hope this will be the closest I'll ever come to giving anyone a chance of calling me an isolationist.)

For a long time, it is more likely that we shall be able to listen in than to participate in a really active manner. There may be among you, however, those who feel—and perhaps rightly—that I have been much too conservative. We should not even be satisfied with trying to get to the center of our galaxy. Why not think about getting to the nearest neighboring galaxy, to Andromeda? It takes light about two million years to go this distance; and Einstein says we can't do it faster. And this time really seems to be just a little bit too long. But I don't want to be deterred—I want to think of the possibility anyway. And fortunately, Einstein gives me a chance to do this; he gives me some hope that one of us actually might get to Andromeda. Even though the present engineers and their children and grandchildren to the seventh generation are clumsy enough not to be able to do it, still the physicists tell us that it might be done. And without prolonging human life beyond any reasonable length. Now, I will explain this; but in order to explain it, it will be necessary for me first to give a brief review of Einstein's relativity. The complicated organic chemistry is gone—the simplicity of physics will appear.

We are going to consider two events taking place in different locations and at different times. The relation between these two events can be characterized by two numbers. One is the distance between these two events which I shall call R; the other is the time difference between these two events which I shall call t. It's uncomfortable to compare distance and time; therefore I will use instead of the time, a distance— that distance which light could have covered in the time t. That distance is equal to the light velocity c multiplied by the time t. How will any individual with some common sense think about distance R and about the time difference t? The time difference will appear to every observer as the same. For instance, suppose that I leave San Francisco and arrive in Los Angeles 1½ hours later. All observers should agree, every reasonable person will say, that the time difference has been just what it was—an hour and a half. But as to the distance, the various observers should reasonably disagree. People staying on the ground will say that the two events of my departure and arrival were occurring at a distance some 400 miles apart, which is the distance between San Francisco and Los Angeles. I will maintain that the two events happened in the same position; namely, in the airplane in which I was present. This is my system, or was while I was sitting in it. Still somebody else, taking a slightly more general view, standing outside the earth and, let us say, in the vicinity of the sun, will say that the distance was much greater than 400 miles, because between my departure and arrival the earth itself has run a very considerable distance, much more than 400 miles. So for various observers, R will have various values, but t, the time interval, will appear to have the same value. This was what seemed to be true up to 1905.

In 1905 Einstein discovered—and there is no doubt that he's right— that this is not true. However, it is approximately true in cases where the distance is relatively small and ct is relatively large. The example of my travel between San Francisco and Los Angeles is such a case. But if you really look at things precisely, it turns out that the time interval is no more definitely given than the space interval. It too will appear to have different values for different observers who differ in no more than their relative speeds. And each of these observers has the full right to say that his is the right observation; no one can say that his observation is more valid than another person's.

Now, so far, I have credited Einstein only to the extent that he exposed an old, imperfect idea. The real accomplishment in the work of

Einstein was that instead of this old idea he invented and proved a new one, which is accurate and correct. The time difference does not remain the same but another quantity, which I shall call Q, does remain the same for all observers. Here is how we determine the quantity Q. Take the distance ct which light could have covered during the time difference which one has observed. Multiply it by itself, which gives $(ct)^2$. Also, take the other distance R and multiply it by itself, which gives R^2. Then, subtract one from the other and we get $(ct)^2 - R^2 = Q$. This is all the mathematics I need. Einstein's assertion is that this quantity Q will remain the same for every observer.

No one can go faster than light—no one really can go even as fast as light—but if all the engineers will work sufficiently hard and give us the right propulsion means and the right vehicle, we can go almost as fast as light; and will you please assume that this will be the case. I predict that this will not happen in my lifetime. Within a hundred years, who knows? Within 500 years, practically everything that is impossible must have happened. I have an idea how we could do it. We have not yet discovered anything that we would call antimatter. But we know that matter consists of electrons, protons, and neutrons, and we have discovered the antielectron, which is called the positron, also the antiproton and the antineutron. Out of these we can surely compose antimatter. And antimatter has the beautiful property that when it unites with matter, both are transformed into pure energy—into radiation. If only we could have a little antimatter (it would be difficult to make), and then trickle by trickle let it unite with matter, the resulting energy could help us to go almost as fast as light. There is one little difficulty (even beyond the question of how to make antimatter), and that is how to contain it; because once it touches matter, there it goes. But I wouldn't worry about it too much. We might somehow find a real ingenious scheme, a very stable and reliable scheme, to hold antimatter by some sort of arrangement of magnetic fields so that it doesn't touch matter and so that it isn't let out too fast. It's merely an engineering detail. So let us assume that this problem will have been solved in the year 2460 and off I go to Andromeda.

Now, let us say that I go almost as fast as light. Therefore, what will you, an observer from the earth, see? You will see that I am heading toward Andromeda almost as fast as light and that I will take just a very little more time than light would have taken: two million years and a very little more. So t will be two million years and a little more, and the distance ct will be a little more than two million light-years. The distance R, on the other hand, will be, let us say, just two million light-years. The difference between the two huge quantities $(ct)^2$ and R^2 will be quite small, since I have gone almost as fast as light. This is what you, the earth observer, sees. But what will I, the traveler, see? I sit in the rocket as I sat in the airplane from San Francisco to Los Angeles. And I will say that my arrival and my departure have taken place in the same location; namely, in this rocket ship behind the controls. As far as I can see, R, the distance between the two events of my departure and my arrival, is zero. But the difference $Q = (ct)^2 - R^2$ must be the same for me, the traveler, as for you, the observer. And since the difference Q is small for you, it must be small for me. And, therefore, the time that has passed must appear short to me. To you it has seemed as more than two million years. To me it may seem as 20 years.

Well, when I get to Andromeda I will be faced with some real dangers. I will mention only one. Here in our galaxy we have good information that if there is any antimatter around it's not much, but for all we know, the Andromeda galaxy may be made completely out of antimatter. So that my first experience in arriving at Andromeda might be to be annihilated. Actually, it's not quite that bad. As long as I stay out in interplanetary or interstellar space where I meet only a few antiatoms, I won't get annihilated so fast. At any rate, to get into an antigalaxy is something that gives me nightmares.

Let us say that Andromeda is a decent galaxy, composed of old-fashioned matter, and I stay there for 10 years exploring around and then I turn back and come home. Now you would imagine that for my achievement I might get a ticker-tape reception in New York. Far from it. I will be 50 years older, but the earth will have become more than 4 million years older. All my friends will be dead; nobody will speak my language, neither English nor Hungarian; my notes might be deciphered slowly by the scientists. A new race will live which we might consider as a strange and horrible new species, but which in reality will be far superior to us and much better. And what they will do with me—a specimen from an old, fabulous, unreasonable, extinct race—is obvious. I will be put in a zoo.

Index